Food Nutrition and Health

Food Nutrition and Health

FERGUS M. CLYDESDALE
and
FREDERICK J. FRANCIS

Department of Food Science and Nutrition
University of Massachusetts
Amherst, Massachusetts

AVI Publishing Company, Inc. Westport, Connecticut

Copyright 1985 by
THE AVI PUBLISHING COMPANY, INC.
250 Post Road East
P.O. Box 831
Westport, Connecticut 06881

Library of Congress Cataloging in Publication Data

Clydesdale, F. M.
Food, nutrition, and health.

Bibliography: p.
Includes index.
1. Food. 2. Nutrition. 3. Food supply.
I. Francis, F. J. (Frederick John), 1921-
II. Title.
TX354.C57 1985 613.2 85-18650
ISBN 0-87055-507-3

Printed in the United States of America
B C D 432109876

Contents

Preface

Dramatic cultural changes have occurred in the areas of food, nutrition, and health in the United States. Today, the clarion call is for fitness with "trim-muscular" in and "skinny-pale" out. The "me generation" has turned into a robust health seeking "we generation," with emphasis on group participation in an ever-increasing array of health clubs. Combined with this renewed interest in fitness is an increasing acceptance of technology, which has resulted in the expectation of a high quality of life through the use of technology rather than through its banishment as was the case in the late 1960s and 1970s. Thus, we see the use of individualized computer programs for diet, exercise, and improvement of athletic performance through motion analysis of the event.

Aging has become an accepted phenomenon and the long fruitless search for perpetual youth seems over. Old is beautiful as long as it is associated with the trim and robust look of other age groups. This is due to the changing demographics of the United States, as well as recognition of the simple fact that age is not a drawback in achievement levels in most areas of our society.

These changes, which are in many respects the antithesis of the beliefs of the 1970s have led us to write this book. We are going to attempt to use the same style of communication we used in our previous book, "Food, Nutrition, and You" but the focus and content is quite different. We will once again attempt to produce the facts and to point out the frauds. The fads and frauds still exist, they have simply adjusted from the needs of the 1970s to the needs of the 1980s.

Intertwined with these changes is a worldwide concern for peace and prosperity. This, we believe, will depend upon the ability of the world to produce enough food as much as it will on arms limitation discussions, since a well fed population is less likely to go to war.

Therefore, we have devoted a segment of the book to the problem of food supply because it would be provincial to deal only with the food concerns of affluent countries. It is our hope that we offer a unique combination of information written such that the nonscientist may both enjoy and benefit from it. It is aimed at those who wish health and fitness in a well-fed world community and are willing to accept long-term goals rather than short-term wishes.

Fergus M. Clydesdale
F. Jack Francis

I

Nutrition and Food Safety in the United States

Part I of this book will deal with nutrition, diet, health, disease, and food safety in the United States. The food-related health problems of the more affluent western world are quite different than of the developing countries where hunger and malnutrition are a common part of everyday life. To put the seriousness of this latter situation into perspective, the President's Commission on World Hunger reported the following facts.

In the next 60 seconds, 234 babies will be born
- 136 in Asia
- 41 in Africa
- 23 in Latin America
- 34 in the rest of the world

23 of these 234 will die before age 1
- 6 in Africa versus 0.01 in North America
- 2 in Latin America versus 0.25 in Europe

34 more will die before age 15 and 50 to 75% of all these deaths can be attributed to a combination of malnutrition and infectious diseases.

The production of more food, its preservation and distribution, would certainly go a long way in alleviating these problems. However, this would only produce short-term results if population control and energy production were not also considered.

There has been little concern over population control and food supply since the crisis situation of the early 1970s, until the recent Ethiopian catastrophy, which clearly indicated that a few years of worldwide crop failures could have any of us on the brink of disaster.

Census figures have shown a dramatic increase recently, with the largest annual worldwide increase in history noted in 1982. Eighty-two million people were added to the world's population, producing a total figure of 4.7 billion people. Coupled with these astonishing numbers is the ever-present problem of energy scarcity which, if it becomes worse, could have a devastating effect on the ability of the large food-producing countries to continue their present rate of production.

In the United States, a transition in food production has occurred. Early in its history, the major food problem was scarcity, as in the rest of the world. However, there has been a revolution in agricultural technology and production, which has provided abundance. This factor along with advances in public health, medicine, and the social and physical environment in which we live has resulted in an ever-increasing life span. In 1900, the mortality rate was about 2,000 per 100,000. That rate has now been reduced by several hundred percent and is still falling. In 1950, it was 841.5 per 100,000, in 1970 it shrunk to 714.3 and today, it is around 590 deaths per 100,000 population (National Center for Health Statistics, 1982).

The following health-related developments have occurred:

- Life expectancy of the newborn climbed to a record 73.3 years, an increase of 2.7 years since 1970.
- The death rate from heart disease, though still the nation's primary killer, dropped nearly 20%. The death rate from stroke declined 33%.
- Fewer people under age 49 are dying of cancer—the result of fewer cases of lung cancer in younger men, and advances in treatment of breast cancer in younger women, Hodgkin's disease, and childhood leukemia.
- The incidence of childhood infectious diseases has fallen so low that it is hoped that measles soon will be eradicated as an indigenous disease. (However, a recent outbreak in Massachusetts might cause some problems in eradication.)

However, though we have the potential to be the best fed and most comfortable people in history, we tend to abuse that potential by abusing our bodies. We eat too much of everything, including alcohol. Satisfactory health care, nutrition education, drug and alcohol abuse, and smoking are still major problems in our society. In fact, the same report that produced the preceding good news also lists some bad news.

- The death rate is rising for 40 million Americans of ages 15 to 24 as a result of accidents, violence, and alcohol and drug abuse.

- Deaths from motor-vehicle accidents, after declining in the mid-1970s are up again. Traffic-death rates rose 4% in 1977 and 7% in 1978.
- Some experts fear that there may be too much unnecessary surgery. One concern: 15% of all births are now by Caesarean section—triple the proportion of a decade ago.
- The nation's health bill now exceeds 212 billion a year. Nursing-home costs have risen 20% each year since 1967.
- Blacks and the poor have higher rates of illness—and stay sick longer—than do whites and the rich. The mortality rate for black infants is almost twice the rate for whites. Poor families see dentists only half as often as people with higher incomes and have more problems with teeth.

Even with these minuses, it is obvious that we are becoming progressively healthier and, it is hoped, this trend will continue. However, care must be taken in the allotment of priorities to health. Nutrition is important, but it is not a national eraser that is going to wipe out disease. Certainly, food and fitness are items of the utmost value, but the individual must realize that both genetic components and life style other than food and fitness contribute greatly to disease. Personal responsibility cannot dictate genetic inheritance even though the key to living to a ripe old age is probably involved with choosing the right grandparents! What can we do?

It is obvious that one cannot escape death. This, of course, is unfortunate, but the best we can hope for is to escape premature death, that is, death between the ages of 25 and 65. One suggestion (Nelson 1982) for the individual is to focus on changes that might reduce the risk of death at an untimely age by first looking at the most common causes of premature death and then calculating the number of deaths attributable to known, and controllable, environmental factors.

Mortality statistics for adults between the ages of 25 and 65 for the year 1978 show that the ten leading causes of death were

Heart disease (all types)	165,731
Cancer	150,981
Accidents (motor vehicles and others)	43,798
Cerebrovascular disease	24,956
Cirrhosis of the liver	21,135
Suicide	17,227
Homicide	12,823

Diabetes mellitus	9,274
Influenza and pneumonia	9,073
Bronchitis, emphysema, and asthma	5,649
Total	460,647

According to current medical evidence, at the very least the effects of the top five causes of premature death could be reduced. The most common cause of premature death is heart disease, of which coronary heart disease is by far the most prevalent. We know that about one-third of coronary heart disease deaths are caused by cigarette smoking (46,600 deaths in 1978). At least an additional 8,800 heart disease deaths are related to high blood pressure (a risk factor for coronary and hypertensive heart diseases). Some aspects of diet may be risk factors for heart disease. Obesity, specifically, increases the risk of high blood pressure, which may lead to heart disease. (However, numerical estimation of deaths attributable to diet or obesity is not yet possible.)

High blood pressure is also an important risk factor for cerebrovascular disease, the fourth leading cause of premature death, but again, no quantitative estimate of attributable deaths for it can be made.

One-third of all cancer deaths were caused by cigarette smoking as well (50,300 in 1978).

At least half of the liver cirrhosis deaths—around 10,500—could have been prevented by moderating alcohol use and improving diet. And 45% of fatal motor vehicle accidents in the 25–65 age group were caused by intoxicated drivers (some 9700 deaths in 1978).

The tenth leading cause of premature death was chronic lung disease. Seventy-five percent of the emphysema and bronchitis deaths were caused by cigarette smoking—around 3600 deaths.

Adding these up, we find that smoking caused more than 100,500 premature deaths, alcohol abuse another 20,000, and high blood pressure was related to more than 8800 premature deaths (and these are extremely conservative estimates). The figures total 130,000 deaths, which is approximately one-fourth of those who died of the top ten causes of premature adult death in 1978.

Interestingly, the statistics provided above by Nelson (1982) are extremely conservative for deaths caused by drunk driving. *Newsweek* (Anon. 1980) quotes 1980 National Safety Council Figures as 26,300 deaths per year due to drunk driving, accounting for fully one-half of all auto fatalities and killing far more Americans per year than any other accidents. Thus, another 16,600 premature deaths over and above Nelson's (1982) estimate could be prevented. In addition, these figures do not include drug use and abuse. We have very little data on the effects of these compounds, but surely they must contribute to a great extent to accidental death.

We could prevent over one-fourth of premature deaths simply by changing our lifestyles. This would not require research dollars, government regulation, government spending, or prohibition of any food or food group. It seems this decision is up to the individual and to how much each person is willing to change.

On the worldwide basis, the answers to prevention of 25% of premature deaths are not as simple. General views on the cause of malnutrition have undergone considerable change over the last two decades. It is now increasingly accepted that there is a relationship between malnutrition, poverty, and economic development (Pellett, 1982). Even the total number of people affected is extremely difficult to estimate, and as Pellett (1982) has pointed out, the statements that so many tens or hundreds of millions presently suffering from hunger or malnutrition do little to help understand or alleviate the problem. Schoen (1983), in an article in *The New York Times,* suggest that hunger is a factor in America again, stating that "Since the 1960's—when America declared war on poverty—malnutrition and the severe diseases associated with it have abated. But today, the subtler forms of hunger that inevitably accompany poverty persist. . ."

There is no question about the fact that some segments of our society can and do benefit from nutritional intervention programs. Some of the problems are due to economics, some to the inability to choose appropriate food, some to a conscious choice of a poor diet due to emotional, psychological, or religious reasons, and probably most due to a combination of many factors.

Studies done at Tulane University and quoted by Burnette (1983) have shown a clear relationship between performance, meals, and nutritional status.

> In a study in 1969–1970, children who received breakfast and lunch at school showed significant improvement in hemoglobin concentrations. . . Children with high hemoglobin levels showed better performance in the Kahn Intelligence Test (KIT) and, in the youngest age group, in simple and disjunctive reaction time tests. . . Similar findings were noted in performance of the associative reaction time test. . .
>
> In studies in 1970–71 . . . children who received breakfast and lunch improved significantly in performance in the disjunctive reaction time and in continued trials of associative reaction time. . . These studies indicated a relationship between mild levels of nutrient deficiency and performance tasks demanding attentiveness and alertness. . . "Nutritional intervention through diet or micronutrient supplementation can improve performance in tasks demanding attentiveness. Micronutrient

supplementation . . . should be a valuable addition to applied health programs in improving nutritional and behavioral status (Smith *et al.*, 1975).

Burnette (1983) also introduces data to support the contributions of the WIC (Special Supplemental Food Program for Women, Infants, and Children) Program in reducing infant mortality. These claim that the infant mortality rate on seven Indian reservations in Montana declined from 31.5 per thousand in 1972 to 16.6 in 1975, following introduction of a supplemental feeding program for pregnant women, infants, and young children, and that Arizona participants in WIC recorded an 81% reduction in anemia, 82% reduction in underweight, and 64% improvement in stature. In Michigan, 30% of the women were anemic before WIC, but only 6% after participation. Anemia among children participating in Oregon was reduced from 13% to 1%. In the Pennsylvania WIC program, the infant death rate was reduced from 10.6% before participation to zero afterwards. Immature birth rates decreased from 12.8% to 1.6%, and pregnancies with complications were reduced from 30.9% to 17.6%.

It seems apparent then that, at least short term, some solutions are available in the United States if food and education are available.

Worldwide, we must attack not only malnutrition, but accompanying problems, such as energy, population control, poverty, and sanitation. Obviously, it is not within our scope to treat all of these, but, as concerned citizens and scientists, we can attempt to alleviate the problem, through education, appropriate technology, agricultural improvement, and the distribution of safe, nutritious, and palatable foods into the hands of the consumer.

BIBLIOGRAPHY

ANON. 1980. The war against drunk drivers. *Newsweek*, Sept. 13, p. 34.

BURNETTE, M. A. 1983. Nutritional needs of the poor in America: challenges for the eighties. VNIS Vitamin Issues III (1), 1. Hoffmann-LaRoche Inc., Nutley, New Jersey.

NATIONAL CENTER FOR HEALTH STATISTICS. 1982. Health, United States, DHHS Pub. No. (PHS) 83-1232, Public Health Service, Washington, DC.

NELSON, N. A. 1982. How best to prevent preventable deaths. ACSH *News Views*. **3**(1), 5.

PELLETT, P. L. 1982. Commentary: Changing concepts on world malnutrition. *Ecol. Food Nutr.* **13,** 115.

SCHOEN, E. 1982. Once again, hunger troubles America. *The New York Times Magazine*, Jan. 2.

SMITH, J. L., JEFFERSON, L., GOLDSMITH, G. A., GOLDSMITH, S. 1975. Prevention of vitamin and mineral deficiencies associated with protein-calorie malnutrition. *In* Protein Calorie Malnutrition, R. E. Olsen (Editor). Academic Press, New York.

1

The Search for Health: Nutritional Considerations

Recently, it was reported in the magazine *Psychology Today* that a "cultural hypochondria" is sweeping the country. They concluded that ". . . like so many Gatsbys in jogging suits, Americans are chasing an ideal of perfect health. Many of us, in fact, now spend more time thinking about our health than about love or money." Obviously, this is a noble ideal, and, on the surface, personal responsibility for our own well being is a refreshing notion. Unfortunately, the end does not justify the means, since many consumers choose to jump at the slightest hint of a health benefit promised by any dietary manipulation that does not involve a major change in lifestyle. They want to know how food can make them stronger, happier, more creative, less prone to illness, and sexually capable, to a degree unheard of in most societies. That is, they want something that will prevent disease or work miracles with a minimum of effort. Who can blame them? This is an ideal we all would accept with open arms, but unfortunately life and health is not all that simple.

Our preoccupation with health is a reasonably new phenomenon, and in part is a result of living in such an affluent age that we have time to worry about such things, but the search for easy remedies has a long historical tradition. Magic, superstition, and myths are recorded in our earliest beginnings. There was always someone searching for the fountain of youth.

In the United States one of the first discoveries of perpetual good health was made by Dr. Elisha Perkins in 1796.

Surprisingly enough, it was quite simple. Dr. Perkins fabricated two metal rods each 3 in. long and each of specially compounded materials. By placing the tractors—as they came to be known—over that area of the body suffering from pain, heat, or disease of some kind, then drawing the rods out to one of the extremities, Dr. Perkins could remove the illness.

Within a matter of weeks Dr. Perkins found that his own healing hands were not needed to guide the tractors. Anyone could do it. Quite understandably, Americans rushed to buy tractors from their benefactor. No price was too large for such a boon. It is recorded, though not well documented, that one Virginian sold his home and accepted tractors enough to total the entire value of his estate. Elisha Perkins had become the first great American quack.

Of course, action was taken by the authorities at once. President George Washington immediately bought tractors for his entire family. And Supreme Court Chief Justice Oliver Ellsworth, after calm judicial reflection, took an even more foresighted course. Having purchased tractors from Perkins and trying them, he introduced the good doctor, with the highest recommendations, to his successor, Chief Justice John Marshall. Medical historians are convinced that Perkins was sincere. To this conclusion they adduce the evidence that during the great yellow fever epidemic that struck New York City in 1799, Dr. Perkins rushed to the very heart of the plague and tended the stricken. Seeing that the fever racked the entire body of its victims, he realized that tractors alone were not enough. Vinegar, quaffed in judicious combination with the magical tractors, he was certain, was the answer. Confidently, fearlessly, he applied his cure. Three weeks later Dr. Elisha Perkins was dead of yellow fever. In so dying, Perkins at once advanced and refuted the first important food fallacy. He also left the heritage of the tractors to his son, Benjamin, who was a quack of the most vicious sort. Cynically, he plied his tractor trade in both the new world and the old, was honored wherever he went and in building a fortune never for a moment gave credence to his father's medical faith. This is a pattern which we shall see again and again.

[Is this story merely a] wry vestige of a distant past? Not quite. Recall Dr. Perkins' finest hour, his ultimate discovery—vinegar to cure yellow fever. Recall Dr. Perkins' vinegar and then consider "folk medicine," by Dr. DeForest C. Jarvis. Dr. Jarvis is said to have sold over half a million books. Not surprising when one understands that Dr. Jarvis offers relief from headache, arthritis, diabetes, vague physical weakness, and other scourges. And his method is so easy. One need only know the proper combinations and dosages of two monumental nostrums. One of these is honey. And the other—shades of Elisha Perkins—is vinegar" (Deutsch 1967).

We did it then, and we are doing it again today, looking for the easy cure that will make us feel better, lose weight, increase our sexual potency, and prevent disease.

Unfortunately there is no single, simple action that will provide good health. Good health depends on some genetic luck, such as having the right grandparents, a strong personal commitment, and the good fortune to live in a part of the world where sanitation, shelter, medical care, food, and water are all available. In the United States, Canada, and much of western Europe, these conditions do exist, and people have the potential to live long and productive lives. However, as pointed out in the introduction, this demands lifestyle changes and a long-term commitment. Obviously, the cessation of smoking and moderation in alcohol use are of prime importance. Exercise and weight control along with an appropriate diet are critical concerns. Balance and harmony in life with a well-nurtured physical, mental, and spiritual basis are all keys to good health. This really is the "wellness" or "holistic" approach, to use the current buzz words, for good health. There is nothing mysterious about it. Commitment, some discipline, a good knowledge base, and moderation will not only reap a longer life but a more joyous life. Such a life allows more freedom in living and it begins with you, the reader. A personal commitment is involved, in that *you* change, that *you* take responsibility, and that *you* enjoy the benefits. Illness and disease are not preventable; they are a part of life, striking randomly. But, you can minimize your risk of disease by taking the appropriate actions. This chapter will deal with some of those actions that are concerned with nutrition and fitness.

DIGESTION AND ABSORPTION

Nutrition is obviously an important component of the "wellness" approach to good health. After all, food provides nutrients, which in turn supply the fuel, construction materials, blueprints, and machinery to maintain the body and to make it run. The body is really a remarkably designed machine from a physical point of view. Unlike a car, which rusts and falls apart, it continually rebuilds itself. Every cell in the body is in a dynamic and ever-changing flux, constantly being rebuilt. Just consider that the cells that make up the lining of the small intestine are lost by extrusion into the intestine at a rate of 20 to 50 million cells per minute! It is indeed fortunate that the food we eat supplies all the necessary chemicals to replace these cells at the same rate.

What are the chemicals that food provides? This might best be answered by looking at the chemical composition of the human body. The largest single component of the body is water, which is made from two atoms of hydrogen and one atom of oxygen. These two elements along with nitrogen and carbon in different combinations also form fat, carbohydrates, and proteins in the body. In fact, if we consider our total body makeup we find that about 98% of our weight would come from compounds formed from carbon, oxygen, hydrogen, and nitrogen. The remaining 2% of our body is made up of many other diverse elements, some of which we know to be essential and others we are not so sure about. However, we do know that all these elements and atoms can combine to form the various compounds that make up the human body, including

1. water
2. protein
3. fats
4. carbohydrates
5. minerals
6. vitamins

The digestive tract, which is in effect a hollow tube extends through the body. At one end of this tube is the mouth, where we ingest food; at the other end is the anus, where we excrete waste materials. As food travels down this long tube, we mix, digest, absorb, and assimilate the chemicals we require to maintain our body; the food ingredients we do not require, we excrete.

This process is shown schematically in Fig. 1.1, where the intestine has been straightened out for simplicity rather than showing it in coils as it actually exists in the abdominal cavity. The objective of the entire process of digestion is to break food down into smaller and smaller compounds until a size is reached small enough for such particles to pass through the intestinal wall in ways similar to the chemicals in tea passing through a tea bag when it is placed in a hot solution. Remember that we cannot absorb a whole apple or an orange, but only the components of these foods which, in fact, are very similar and consist mainly of sugar and water along with smaller amounts of proteins, vitamins, minerals, fiber, and trace amounts of other chemicals.

Treated in such a way, the human body seems a rather simple creation, and indeed it is. Materially, it costs very little, even in these days of inflated prices: for about a hundred dollars we could purchase from the local supermarket, drug stores, and chemical houses all the ingredients that make up the body. However, before we deflate your ego totally, we should note that chemicals of the human body are organized in the form of cells. Cells are tiny, intricate bundles of life that are responsible for vast numbers of chemical reactions. They contain such things as the genes and chromosomes that determine our heredity and the components of the nervous system that allow us to think and act as

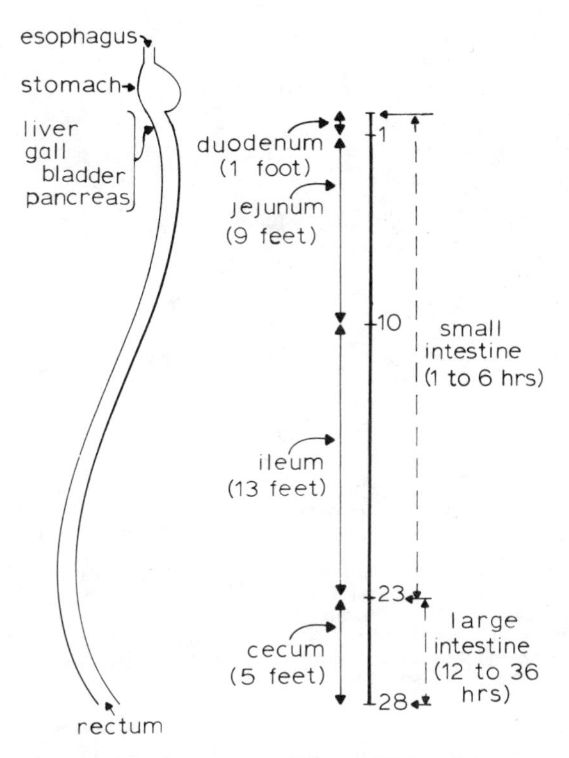

Fig. 1.1. A schematic representation of the digestive tract commonly known as the gastrointestinal system.

rational beings. Each of the billions of cells that we are made up of is like a tiny city containing factories that assemble parts, energy plants, blueprints, and engineers who take the blueprints off the drafting table and put them to use in the factories to create the components that we need to live. This chapter, however, is not aimed at even a simple discussion of molecular biology, and we will now leave this discussion in order to evaluate further what our body needs from the simple materials known as food.

RECOMMENDED DIETARY ALLOWANCES—THE RDA

The next obvious question is "How much of the nutrients do we need?" Is more always better or is there an optimum intake below which deficiency might result and above which toxicity could be caused? In-

TABLE 1.1. Food and Nutrition Board, National Academy of Sciences-National Research

	Age (years)	Weight (kg)	Weight (lb)	Height (cm)	Height (in)	Energy Needs (with range) (kcal)		(MJ)	Protein (g)	Fat-Soluble Vitamins: Vitamin A (μg RE)[b]	Vitamin D (μg)[c]	Vitamin E (mg α-TE)[d]
Infants	0.0–0.5	6	13	60	24	kg × 115	(95–145)	kg × 0.48	kg × 2.2	420	10	3
	0.5–1.0	9	20	71	28	kg × 105	(80–135)	kg × 0.44	kg ×2.0	400	10	4
Children	1–3	13	29	90	35	1300	(900–1800)	5.5	23	400	10	5
	4–6	20	44	112	44	1700	(1300–2300)	7.1	30	500	10	6
	7–10	28	62	132	52	2400	(1650–3300)	10.1	34	700	10	7
Males	11–14	45	99	157	62	2700	(2000–3700)	11.3	45	1000	10	8
	15–18	66	145	176	69	2800	(2100–3900)	11.8	56	1000	10	10
	19–22	70	154	177	70	2900	(2500–3300)	12.2	56	1000	7.5	10
	23–50	70	154	178	70	2700	(2300–3100)	11.3	56	1000	5	10
	51–75	70	154	178	70	2400	(2000–2800)	10.1	56	1000	5	10
Females	11–14	46	101	157	62	2200	(1500–3000)	9.2	46	800	10	8
	15–18	55	120	163	64	2100	(1200–3000)	8.8	46	800	10	8
	19–22	55	120	163	64	2100	(1700–2500)	8.8	44	800	7.5	8
	23–50	55	120	163	64	2000	(1600–2400)	8.4	44	800	5	8
	51–75	55	120	163	64	1800	(1400–2200)	7.6	44	800	5	8
Pregnant						+300			+30	+200	+5	+2
Lactating						+500			+20	+400	+5	+3

[a] The allowances are intended to provide for individual variations among most normal persons as they live in the United States under usual environmental stresses. Diets should be based on a variety of common foods in order to provide other nutrients for which human requirements have been less well defined.
[b] Retinol equivalents. 1 retinol equivalent = 1 μg retinol or 6 μg β-cartoene.
[c] As cholecalciferol. 10 μg cholecalciferol = 400IU of vitamin D.
[d] α-Tocopherol equivalents. 1 mg *d*-α-tocopherol = 1 α-TE.
[e] 1 NE (niacin equivalent) is equal to 1 mg of niacin or 60 mg of dietary tryptophan.
[f] The folacin allowances refer to dietary sources as determined by *Lactobacillus casei* assay after treatment with enzymes (conjugases) to make polyglutamyl forms of the vitamin available to the test organism.

deed, there is an optimum amount. Deficiencies do occur and death can result from too much of some of the nutrients. However, how can we determine the exact requirement for nutrients for each individual since we are all different? The answer is that the exact requirement cannot be calculated; we would have to study each individual on the earth in a hospital setting to find such requirements. Therefore we must study whole populations and recommend standards that fit the entire population while remembering that individual needs within that population will vary.

Therefore, the Food and Nutrition Board of the National Research Council of the National Academy of Science was asked to set Recommended Dietary Allowances (RDA) for populations of men, women, pregnant and lactating women, and children of various ages. These are sound guidelines for nutrient requirements; the most recent recommendations are shown in Table 1.1. These recommendations are not individual requirements but are "recommendations for the average daily amounts of nutrients population groups should consume over a period of time" (Recommended Dietary Allowances 1980). As such, they are

Council Recommended Daily Dietary Allowances, revised 1980.[a]

Water-Soluble Vitamins							Minerals					
Vitamin C (mg)	Thiamin (mg)	Riboflavin (mg)	Niacin (mg NE)[c]	Vitamin B_6 (mg)	Folacin[f] (μg)	Vitamin B_{12} (μg)	Calcium (mg)	Phosphorus (mg)	Magnesium (mg)	Iron (mg)	Zinc (mg)	Iodine (μg)
35	0.3	0.4	6	0.3	30	0.5	360	240	50	10	3	40
35	0.5	0.6	8	0.6	45	1.5	540	360	70	15	5	50
45	0.7	0.8	9	0.9	100	2.0	800	800	150	15	10	70
45	0.9	1.0	11	1.3	200	2.5	800	800	200	10	10	90
45	1.2	1.4	16	1.6	300	3.0	800	800	250	10	10	120
50	1.4	1.6	18	1.8	400	3.0	1200	1200	350	18	15	150
60	1.4	1.7	18	2.0	400	3.0	1200	1200	400	18	15	150
60	1.5	1.7	19	2.2	400	3.0	800	800	350	10	15	150
60	1.4	1.6	18	2.2	400	3.0	800	800	350	10	15	150
60	1.2	1.4	16	2.2	400	3.0	800	800	350	10	15	150
50	1.1	1.3	15	1.8	400	3.0	1200	1200	300	18	15	150
60	1.1	1.3	14	2.0	400	3.0	1200	1200	300	18	15	150
60	1.1	1.3	14	2.0	400	3.0	800	800	300	18	15	150
60	1.0	1.2	13	2.0	400	3.0	800	800	300	18	15	150
60	1.0	1.2	13	2.0	400	3.0	800	800	300	10	15	150
+20	+0.4	+0.3	+2	+0.6	+400	+1.0	+400	+400	+150	*h*	+5	+25
+40	+0.5	+0.5	+5	+0.5	+100	+1.0	+400	+400	+150	*h*	+10	+50

[g] The recommended dietary allowance for vitamin B_{12} in infants is based on average concentration of the vitamin in human milk. The allowances after weaning are based on energy intake (as recommended by the American Academy of Pediatrics) and consideration of other factors, such as intestinal absorption.

[h] The increased requirement during pregnancy cannot be met by the iron content of habitual American diets nor by the existing iron stores of many women; therefore, the use of 30–60 mg of supplemental iron is recommended. Iron needs during lactation are not substantially different from those of nonpregnant women, but continued supplementation of the mother for 2–3 months after parturition is advisable in order to replenish stores depleted by pregnancy.

designed to provide assistance in large-scale foodservice operations so that the nutrients in the food offered each day could be compared to the RDA. They are different than the United States Recommended Dietary Allowance (US RDA), which does not differentiate for age or sex but simply provides one recommendation for each nutrient for everyone. It is used by the Food and Drug Administration in food labeling. This number is usually the largest number for that nutrient from the RDA Table.

It must be remembered that these numbers err on the side of safety, and for nearly everyone represent more than they require unless they are suffering from trauma, burns, or a specific disease state. As Harper (1984), a former chairperson of the Food and Nutrition Board, points out

> Because people differ in size and genetic makeup, their nutrient requirements vary; requirements of individuals for essential nutrients range from about 50% below to 50% above the population average. The RDA, therefore, were set high enough to ensure that, if the quantities of nutrients in the food being served met this standard, they would meet the needs of individuals with the highest requirements. Thus, the amounts of nutrients required by most people will be below the RDA and about half the population should require less than half the RDA. Ob-

viously a dietary standard of this type cannot be used to determine if the intakes of people who are consuming less than the RDA are inadequate. Using the RDA as standards for evaluating the adequacy of nutrient intakes is like setting the standard for height at 7 feet and assuming that all of those under 7 feet have suffered growth retardation.

Therefore if we are consuming each nutrient in an amount approximating the RDA, our intake is likely to be optimum.

Now, let us discuss some of these key nutrients in a little more detail. The major requisites for human life are air, water, and energy. The air we breathe contains oxygen, which is absorbed through the lungs into the blood stream and travels through the *circulatory system* to provide every cell in the body with this oxygen. Both water and food are ingested via the *digestive system* and enter the blood stream from the duodenum where they are also carried to every cell in the body. Water provides a mechanism to solubilize all the chemicals in the body and allows them to interact with one another or be transported in various water-based solutions, such as blood, throughout the body. The food components provide the energy nutrients: carbohydrate, fat, and protein, which react with oxygen in the cells to produce energy and carbon dioxide. We use the energy to function, and the carbon dioxide is carried back to the lungs by the blood and we breathe it out. A truly remarkable system!

WATER

Approximately 75% of the body is water, which is distributed among the fluids inside the cells, the fluids between the cells, and the fluids in the blood. The water within the cells provide a medium in which chemical reactions take place. The function of the water outside the cells is to transport substances and chemicals from one part of the body to another. The third function of the water in the body is to maintain body temperature. When we perspire, water is released through the skin. This water evaporates and thereby cools us, due to what is known as the latent heat of evaporation of water. This cooling effect allows us to maintain the very delicate balance of temperature required to sustain life; we all know the feeling that occurs when we are ill and have a fever.

A fourth function of water is to carry out any by-products or end products of the chemical reactions going on within the body. We should stress again that we are dealing with chemicals and with chemical reactions. These are not scary nor are they unknown. In any chemical reaction, there are quite often by-products or end products that are toxic. Many of these by-products are extracted from the blood by the kidneys and excreted through the urine. The kidneys can handle most toxic substances that are adequately diluted. However, if toxic substances become too concentrated, the kidneys cannot handle them; as a result these substances build up, eventually causing body malfunction and perhaps even death. Such common elements as sodium, which occurs in table salt, can build up and become toxic unless diluted with water. When sodium begins to build up, the organs send a message through the nervous system to the brain, which then informs us that we are thirsty. This is a phenomenon that everyone has undergone after eating a salty meal or otherwise consuming large amounts of salt. Another simple chemical, urea, is formed from the nitrogen in proteinaceous foods. Because of the body's need to dilute urea, we feel thirsty after eating a high-protein meal such as a large steak, a large roast beef dinner, or a large fish dinner. These are but two of many other body chemicals that need to be diluted.

People can live for very long periods of time as long as they ingest sufficient water. Obviously, there are not many people who would wish to volunteer for an experiment designed to find out how long one can live without water; the longest period on record is 17 days. The body requires about 2 quarts of water a day under normal conditions. Under stress conditions—when the temperature is very hot, or we are working very hard—the requirement could be as much as 8–11 quarts per day. This may be obtained from beverages such as soda or milk, but not from alcohol. Alcohol is a diuretic, that is, it causes the body to expel water as urine and therefore can cause us to lose more water than we take in. We also obtain a great deal of our water requirement from food. Vegetables are 90–95% water; fruits are 80–85% water; meat, fish, and poultry can run as high as 70–75%. A healthy body requires water and diets that promote diuretics or limited intakes of water should be avoided.

ENERGY, CALORIES, AND WEIGHT CONTROL

Prior to discussing the individual nutrients that the body requires, we should consider how foods give us the energy to run this intricate machine. Food supplies us with energy, which we express in terms of calories. By scientific definition, a kilocalorie (1000 calories) is the amount of heat (energy) required to raise the temperature of kilogram (kg) of water 1°C. Although in actuality this is the definition of a kilocalorie, due to a long tradition in the nutritional sciences a kilocalorie has become known as a calorie. However, this definition does not really allow us to conceptualize the function of the calorie. To do this, it might be simpler to think of a calorie as a unit of energy supplying the body with a certain amount of power, just as a gallon of gas powers a car. Different nutrients give us different amounts of calories, based on the ratio of carbon to hydrogen to oxygen they contain. In general, a gram of protein supplies us with 4 calories; a gram of fat, 9 calories; and a gram of carbohydrates 4 calories. We can see that a given weight of fat supplies us with more than twice the energy of the same amount of protein or carbohydrates. This is why fats are considered "fattening." If we take in more energy than we require, this energy is stored in the body as fat and in time results inweight gain and obesity.

The total energy requirement of an individual can be divided into two types. First, the body needs energy in order to perform the work involved in the process of living—the beating of the heart, breathing, the activity of the glands, and keeping the muscles in a normal state of tension. This involuntary work is necessary to life and continues 24 hours a day throughout life. The rate at which calories are burned in the performance of this work remains reasonably constant over long periods of time, and is called the basal metabolic rate (BMR). The second type of calorie requirement depends on how active the individual is in his or her work and leisure activities. The person who holds a sedentary job and has no vigorous leisure pursuits will certainly require far less energy in the form of calories than one who has a physically demanding job or who pursues active recreations. We can say, then, that the first demand for calories is in the form of involuntary energy requirement; the second demand is in the form of voluntary energy requirement—that is, requirements besides those that sustain life.

The BMR of a normal individual is constant; any large variation is a sign of poor health. One's BMR may be determined by instruments that measure oxygen utilization, respiratory capacity, and so on, or it may be determined by certain simple formulas. For instance, Deutsch (1976) has suggested that an adult male calculate his BMR by multiplying his weight by 12 and an adult female by multiplying by 11.

The BMR of juveniles is much higher than that of adults, per unit of weight. This is due to the energy required for growth and to the different body composition of juveniles. The BMR is highest per pound of body weight at about age one, and then drops somewhat with slight variations until the individual stops growing.

Having discussed the two basic requirements for calories, we can begin to understand the problem of weight control. Weight control has become almost an obsession with the American public, and for good reason. Obesity is associated with an increased risk of contracting many major illnesses such as heart disease, cancer, and diabetes. It should be stressed that this conclusion comes from epidemiological data, that is, data obtained by correlating different occurrences in large population groups. Thus, in a large population, more obese people suffer from these diseases than those who are not. This does not mean that obesity causes these diseases, but it does mean that obesity correlates with a greater prevalence of these diseases indicating that those who are obese are at greater risk of contracting them.

This, unfortunately, is not the major reason for the interest in weight control. As is often the case, the American public responds better to cosmetic appeal than to disease control, and this is the reason why, year after year, Americans spend billions of dollars and grasp at unbelievable plans for weight reduction, including gadgets that promise to melt fat while watching television. They believe that a once-a-week round of golf, riding in a cart, an evening of bowling while sipping a glass of beer, or a short swim and a long sunbath will keep them trim, vigorous, and youthful. They pound and rub and roll their fat in an effort to "break it down" or move it from hips to bust. They look for instant and lasting rewards from sporadic efforts to tone their sagging muscles. And when they discover a jogging session or a set of tennis leaves them sore and strained, they become disenchanted; they feel betrayed by vigorous exercise and head for the steam or sauna bath. They expect the impossible from exercise because they do not understand what the possible is. In addition, they believe in diets which are silly at best and dangerous at worst.

How much energy does a person require? We know that our body requires a certain number of calories to sustain itself—that is, to maintain its basal metabolic rate. This amount is about 1400 for an average woman weighing about 121 pounds, and about 1600 for an average man weighing about 143 pounds. Any excess calories that we consume must be burned up by the work we do or the recreational activities we engage in; otherwise, this excess energy is stored as fat.

It is a simple and easily remembered fact that 3,500 excess calories will produce 1 lb of fat. These 3,500 excess calories typically accumulate not in 1 day or 2 but over a period of time. That is, if you consume 500 calories more per day than your body requires (that is, 500 calories more than are required to sustain your BMR plus your voluntary activities), then in 1 week you will gain a pound. Conversely, if you consume 500 calories a day less than what your body needs, you will lose a pound. There are some interesting studies showing that these figures do not hold exactly, and indeed the effect of exercise on the BMR in humans remains a controversial area (Vaselli *et al.*, 1984). However, in all cases "calories do count" and a deficit must be maintained in order to lose weight.

From a subjective point of view, we can attest to the psychological consequences of obesity and the difficulty in attaining weight loss from discussions with many of the approximately 25,000 students taught in our introductory course in Food Science and Nutrition. It seems that the motivation and trauma are at least equal and often greater in those attempting to lose a relatively few pounds as that in those who are truly obese. This is due in part to the fact that many young women have a physical self image that is considerably less than the ideal. For whatever reasons, this may lead to eating disorders such as anorexia nervosa. A person (most often women from adolescence to 30 years of age) diagnosed as anorexic is filled with an intense fear of gaining weight even when already emaciated. This condition may ultimately lead to death if intervention in the form of therapy is not undertaken. Bulimia is another disorder. It involves craving food and subsequent gorging, which may also involve purging by means of vomiting, fasting, and the use of laxatives. Bulimics are often aware that their behaviors are unusual and dangerous, but are unable to stop without intervention. Obviously, diagnosis of these disorders should be considered if rapid weight loss and/or weight swings are noted in an individual.

The complex interrelationships between appetite, satiety, energy intake, energy requirements, genetics, and environment in obesity are not fully understood. Many factors are thought to be involved, including number and size of fat cells (hyperplastic–hypertrophic obesity due to unusual numbers and size of adipose cells), size of cells (hypertrophic obesity due to unusually enlarged cells), lack of brown fat (and subsequent low energy thermogenesis), elevated levels of the enzyme lipoprotein lipase, and a high set point, where it is postulated that the body demands more calories and less exercise than normal. However, as interesting as all these factors are, the fact remains that in most people with a safe level of calories in a balanced diet, weight loss and weight gain are controlled by the laws of thermodynamics. Energy intake must equal output or an imbalance in one direction or the other occurs.

Having established energy requirements, an individual trying to lose weight should be told that calorie intake should be restricted so that a deficit of 500–1000 calories per day is achieved. Since 1 pound is lost for every 3500 calorie deficit, a weight loss of 1–2 pounds per week results. This should be achieved with no less than 800–1000 calories consumed per day for safety, which means daily caloric need must be at least 1300 calories (1300 − 800 = 500 calorie deficit per day) for a loss of 1 pound a week, and 1800 calories (1800 − 800 = 1000 calorie deficit per day) for 2 pounds per week, which often involves an exercise program.

Having established the thermodynamic basis of weight reduction, it is of some interest to scrutinize the various weight reduction schemes offered in such ubiquitous numbers to the American public. Clydesdale (1984) has suggested that a visitor to this planet would observe that all one had to do to lose weight was to buy a book or magazine with a new diet in it. They might wonder, though, whether any of these books or magazines contained any useful information because if they did there would be no need for a new one every month. The visitors would therefore conclude that weight reduction must involve the act of purchasing the books and not the act of reading them or following their diets. Unfortunately, these visitors, like the consumers who buy these books, would find that just the act of buying them does not work, and often even reading them and attempting to follow their advice is a futile endeavor. The types of diets most often offered vary in complexity and wisdom. They all offer quick, easy weight loss, usually with a gimmick. Some even offer amazing logic, such as that in the Beverly Hills Diet, which says "As long as food is fully digested, fully processed through the body, you will not gain weight. It's only undigested food, food that is "stuck" in your body, for whatever reason, that accumulates and becomes fat." Students have even told us that they lost 10 lb in 2 weeks on a diet that suggested "grab a mate instead of a plate." Unfortunately, the thermodynamics involved require an activity level that even Superman might find difficult to achieve.

Since there are literally thousands of individual diets in the marketplace today, it is impossible to discuss them all. Therefore, we have decided to use a simple scheme suggested by Clydesdale (1984), which groups the available diets into four major categories.

1. Caloric restriction with a balanced food intake
2. Caloric intake controlled through manipulation of diet or use of a "special food" such as a homogenate
3. Caloric intake minimized and a "pill" emphasized
4. An imbalance of carbohydrate, protein, or fat along with either free or restricted calorie intake

Any number of diets would fall within the first category, and as long as nutritional needs are met and the energy balance is reasonable they may be used. The Pritikin program and the recent F-Plan Diet, both of which recommended high fiber intake, fall into the second category, along with any number of diets using special homogenates which can be purchased. The notion that food fiber may prevent obesity results from suggestions that foods high in fiber are filling but contain fewer calories, and that such foods promote rapid passage through the intestine and thus fewer calories are absorbed. However, the utility of a fiber-rich diet in the treatment of obesity has yet to be fully evaluated (Van Itallie, 1980).

A cursory glance at any weekly periodical will show an example of the "magic pill" diet of category 3, which is often simply questionable advertising of category 1 diets. Some of the pills are vitamin supplements with a little sugar, others may be anorectic (appetite suppressing) drugs like phenylpropanolamine, which the FDA has warned may increase blood pressure if misused.

The fourth category is based on the physiological response to dietary restriction. A diet that is low in both carbohydrate and fat is ketogenic (that is, it produces chemicals called ketones) which has led to dangerous health conditions.

A high protein (ketogenic) diet (900–1500 cal/day) does not have any particular advantage over a balanced (non-ketogenic) diet in sparing protein or inducing fat loss. Its major appeal is that its diuretic effect during the first 10 to 14 days provides a striking "apparent weight loss." However, the body soon rehydrates and dieters are amazed that they gained 10 lb over the weekend on a "normal diet." The protein-sparing fast conducted only under strict medical supervision is quite different than the ketogenic diet just described. In the medically supervised fast, caloric restriction is limited to 300–400 calories a day of a high protein supplement. The body adapts, during fasting, after its glucose stores are used and the brain uses ketones from fat rather than glucose made from the amino acids in protein. Thus, lean body mass (protein) may be conserved while weight reduction is achieved. This fast, however, may be hazardous, and it is essential to provide sufficient supplementary potassium to prevent deficiency. A number of deaths have occurred in the United States and Canada due to the use of liquid protein diets (HEW, 1977) apparently caused by ventricular arrhythmias and subsequent heart failure. In spite of this, a weight loss program called the Cambridge Diet was very popular before it went bankrupt. This diet is a product, not a book, which is mixed with water to provide 330 calories a day. It is sold to the customer by a salesperson and not in a store. The American Council on Science and Health (1982) has stated

> During the introductory phase of the Cambridge program, the diet provides only 330 calories a day. This is an extremely low calorie diet. Health authorities, including ACSH, recommend that people not try this type of extreme diet except under close medical supervision. This type of regimen is usually used only as a last resort; it certainly isn't appropriate for people who have only a few pounds to lose.
>
> If you want to use the Cambridge Diet, or any other very low-Caloric regimen, you should do it only with the careful supervision of a physician, who can determine if this type of diet is safe for you, and who can make frequent checks for the dangerous side effects that diets of this type sometimes cause.

Further, the Food and Drug Administration is looking at low calorie protein diets as evidenced by the following statement in Food Chemical News (1984):

> Dieting continues to be a very common and popular activity among Americans of all ages. Protein products subject to the warning regulation continue to be recommended for use and used in very low calorie diets. Moreover, very low calorie diet products that are not subject to the warning regulation have also become very popular. One of these diet products is promoted with a diet plan that is said to provide only 330 Calories per day. x x x There have been reports of serious illnesses and deaths in persons who either were on or had recently completed this diet. Although FDA has not completed its evaluation of the information relating to this diet and is not prepared at this time to propose any action with respect to the diet, the agency is concerned about the continuing popularity of very low calorie diets in general and the possibility that very low calorie protein diets could again become popular.
>
> More important, no substantive scientific evidence was submitted in response to the proposal that in any way alters the agency's conclusion that there is an association between very low calorie protein diets and sudden and unexpected death due to cardiac arrhythmias. The seriousness of this risk in a situation in which the average consumer might be unaware of the risk necessitates a labeling warning about the potential harm associated with the use of protein products in very low calorie diets. . . It is the agency's judgment that the best means to ensure that such harm does not recur is to issue the protein products label warning regulation.

Other diets have different combinations of macronutrients such as the recent book by J. J. Wurtman entitled "The Carbohydrate Craver's Diet," but evidence is very limited as to their efficacy.

Not only should the consumer question the effectiveness of these diets and their potentially overt danger, but they should also look closely at the nutrients provided by such diets. Fisher and La Chance (1984) evaluated the nutritional adequacy of 11 popular weight-loss diets and found that none of them provided all the recommended amounts of the 13 vitamins and minerals studied. Also, they found that the nutrients most often below recommended levels were vitamin B_1 (thiamin), vitamin B_6 (pyridoxine), vitamin B_{12} (cobalamine), calcium, iron, zinc, and magnesium. Therefore, choose with care to assure yourself of nutritional adequacy.

Despite all the weight reduction diets, both good and bad, the social and health pressures, and the use of supportive therapy, the failure rate for loss and maintenance of weight loss has been estimated to be as high as 95%. The reasons for this are extremely complex and we do not pretend to have the answers here. However, a few observations might be pertinent. It seems that dieters respond best when told how difficult weight reduction is. It is not a moral issue, the overweight are not sinful; however, they may not enjoy life as much on a daily basis carrying the extra pounds. Weight loss and diets should not be thought of as interim behavior, which will end when the ideal weight is achieved. Successful dieters have agreed that a committment to a new and permanent life style is necessary. They are not going on a diet, they are changing their way of life! Yearly, not daily, results are important, and the successful program includes exercise in this new way of life. In fact, one of the major causes of obesity in this country is probably the fact that we expend far less energy than we used to in the daily course of events, but we have not cut down our food intake.

Exercise is often said to be ineffective in reducing weight, because in order to lose a pound of fat you must exercise for 7 or 8 hours—longer, if you do not exercise very vigorously. For instance, you must walk 36 hours to lose a pound of fat. This is true, but you could walk one-half hour per day for 10 weeks, a rather pleasant exercise, and lose the same pound of fat. This fact, along with the effect on HDL (a desirable form of cholesterol) and on the ratio of lean to fat in the body, the alleviation of depression, and the maintenance of the BMR under reduced caloric intake it potentially provides, makes exercise an attractive ingredient in a new life style, as will be discussed in a later chapter.

Many of the fad diets and so-called exercise machines and rubberized girdles that claim to remove inches of fat from the user are based on the fact that fat cells in the body contain a high proportion of water. When one uses these rubberized girdles or exercise machines while sitting, there is indeed a dehydration effect and the person's fat cells lose some of the water inside them. Now if you consider that the body contains a great many fat cells, and if each cell is reduced in size due to a loss of water, you can see that an inch or an inch and a half may certainly disappear from a particular part of the anatomy. However, this is simply transitory, and as soon as enough liquid is ingested the so-called loss is replaced immediately.

Some weight-reducing plans advertise a fat loss of as much as 15 lb a week. Think back to the 3,500 calories required to lose 1 lb of fat. Normal, healthy people do not lose fat at that rate. In order to lose 15 lb in a week, you would have to consume 7,500 fewer calories a day. In other words you would have *to need* 7,500 calories per day to begin with. This means that for a week you would have to exercise harder than an athlete training for competition and not eat one calorie. Anyone who knows anything about nutrition should disregard such advertisements.

In order to maintain or lose weight, we must understand the balance between caloric intake and energy expenditure. Our body is just like a bank: if we put more calories into it these calories will accumulate; the only way we can take them out is to spend them by exercising. Therefore, the really healthful method of weight reduction is to cut down the caloric intake and to increase energy expenditure through pleasurable exercise. As Jean Mayer has stated, the faddish way to lose weight should be described as the "rhythm method of girth control."

More and more Americans are leading a more sedentary way of life where their appetites exceed their need. Such individuals have only three choices in regard to weight maintenance: to be hungry all their lives, to become fat, or to exercise more.

It is unwise to undertake a severe diet without a physician's supervision. If weight loss is kept to a pound or two a week, and if a properly balanced diet is maintained, there is not much danger of harming oneself. However, if one undertakes a severe diet or even a moderate one that is unbalanced, anything from a vitamin deficiency to nervous system damage could result.

The bottom line for weight reduction seems to be based on the same advice you will read many times in this book. Reduce calories; eat a little less meat; increase your consumption of fruits, grains, and vegetables; increase your exercise; and remember, reduced weight might not help you to live forever, but it will help you enjoy today.

THE MACRONUTRIENTS: PROTEIN, CARBOHYDRATE, AND FAT

It is fitting that after discussing energy needs of the body we turn to the macronutrients that supply that energy, as defined by the calories they contain. At times one gets the uneasy feeling that the average consumer subconsciously believes in some magic form of energy that does not have any calories. Let us set your mind at rest, there are no such magic nutrients. In fact, the human is limited to carbohydrate, protein, and fat, plus alcohol, as their only sources of energy. Since alcohol should not be considered a healthy or viable source of energy, we shall not discuss it in any detail other than to point out that it supplies about 7 calories per gram, meaning that an average one ounce serving of liquor would provide about 100 calories, an average bottle of beer about 150 calories, and a four ounce glass of wine about 100 calories.

The term "macro" as a prefix simply means "large" and of course "micro" means small. Therefore, the macronutrients are those nutrients we must consume in large amounts to provide energy for our bodies. Each of these nutrients contains carbon, hydrogen, and oxygen, while protein also contains nitrogen, and it is the ratio of carbon to hydrogen to oxygen that determines the number of calories they provide.

Prior to discussing each of these macronutrients individually, it is important to note that as constituents of the food we eat they not only provide energy but also provide a basis for the taste, color, texture, keeping qualities, and overall acceptability of that food. It should never be forgotten that we eat food, not nutrients, and therefore should consider the total role of nutritional compounds in food. In keeping with this concept, note the following definition of food developed by Paul Lachance (1973) of Rutgers University:

Food = known chemicals plus unknown chemicals, which may be modified by
Intentional additives and/or contaminants, which may be modified by
Maturation, storage, processing, preparation, which may be modified by
Digestion, absorption, metabolism, the end products of which may have a
Good, bad, or no effect on body cells and tissues.

We must consider nutrients present, calories present, safety, cost, absorption, digestibility, and acceptability. We need some 50 chemical nutrients to maintain health: water, carbohydrate, fat, protein, 13 vitamins, 17 minerals, several fibers and some trace amounts of other components. Obviously, one food cannot provide all of these, and therefore no one food can provide good health. Therefore, we cannot really speak of a nutritious food since that implies that all the nutrients are present. We must therefore speak of a combination of foods as being nutritious, that is, a nutritious diet. Since the term "nutritious food" is undefinable, how can we differentiate foods? We can do this on the basis of nutrient density, which is defined as the ratio of nutrients contained in a food to calories in that food. Nutrient dense foods have a high ratio of nutrients to calories, and the reverse would be true for foods that are not nutrient dense.

Protein

High protein diets for the weight conscious, protein pills for the body builder, high protein snacks for the child—is it any wonder that consumers in our society are oversold on the wonders of excess protein? The word protein has a magic ring to it in our society, and we are conditioned to believe that more is better. Interestingly, recent surveys have indicated that most Americans, no matter what ethnic group or economic level, ingest more protein than their bodies require. We believe ourselves deficient in protein and feel that somehow protein does not add calories; it only adds "goodness" or "muscle" or "health." This is not to deride the importance of protein, but simply to attempt to place it in proper perspective. Once metabolic demands are satisfied, excess protein is used like calories from carbohydrates, and therefore such excess protein may be classified as being "empty calories."

Proteins are large molecules, which consist of carbon (C), oxygen (O), hydrogen (H), and nitrogen (N). They are constituents of every living cell in the human body. In an adult, three quarters of the residue remaining after water is removed is protein, of this amount one-third is in the form of muscle, one-third in bone and cartilage, one-tenth in skin, and the rest in other tissues and in body fluids. All enzymes are protein in nature. Enzymes are biological catalysts that speed up the rate of the chemical reactions that occur in the body, and in this way they are essential to life. Many hormones are either proteins or protein derivatives. (Hormones are similar to enzymes, but they transport messages in the body.) Materials in the cell that are responsible for the transmission of genetic information and for cell reproduction often occur in combination with proteins as nucleoproteins. Bones, muscles, hair, nails, cell membranes, antibodies and just about everything else in the body is composed at least partly of proteins. Because these tissues regularly need additional proteins for maintenance, the body must have a regular supply of protein in the diet.

Once ingested, proteins are broken down into their component amino acids, which are then resynthesized into new proteins that the body can use. These proteins formed by the body are used for regulating the internal water and acid–base balance, for energy, for building enzymes, antibodies, and some hormones, for growth in children and pregnant women, and also for lactation. Protein is needed also for tissue maintenice or repair in case of injury or blood loss. It is recommended that 56 grams of protein be consumed per day by men over 19 years of age, and 44 grams by women the same age. The recommended dietary allowances for other ages, pregnancy, and lactation vary and may be seen in Table 1.1. It is generally recognized that this figure is a bit generous for our daily needs, but it is nevertheless accepted because a slight surplus of protein is necessary in case of tissue trauma. The body can tolerate as much as 300 grams daily without harm, as long as sufficient water is consumed so that the nitrogenous waste can be disposed of through the kidneys. Remember, when large amounts are ingested and the excess is used for energy, and/or glucose and fatty acid synthesis, the nitrogen has to be diluted so that the kidneys can handle the urea production and so that toxic end products will not accumulate. The large amount of water recommended in some high protein diets is partly for this reason.

But it makes no sense to ingest so much protein that you have to drink more water than usual in order to rid your body of toxic products.

Digestion

The body is unable to absorb proteins intact and therefore must break them down into their component units, called amino acids. The process that breaks down the protein depends on the release of hormones that stimulate the secretion of acids and enzymes in the stomach and other enzymes in the intestine. This combination of acid and enzymes causes the breakdown of the protein into amino acids.

There are some 23 amino acids, which can join together by what is called a "peptide" bond in various combinations to make up most protein. Most protein we eat is broken down to some of these 23 amino acids in the intestinal tract and then absorbed into the blood. We do not absorb any nutrients from the stomach. This makes sense since stomach acid would also be absorbed and cause problems with our internal structure. When material passes from the stomach into the duodenum (Fig. 1.1) the acid is neutralized and it is therefore safe to absorb. In fact most absorption occurs in the small intestine.

Essential Amino Acids

There are eight amino acids that are classified as essential amino acids for adults. These are valine, lysine, threonine, leucine, isoleucine, tryptophan, phenylalanine, and methionine. Histidine, is essential in children and may also be in adults, as is arginine.

These amino acids are termed essential not because they are the only ones needed by the body but because they cannot be synthesized by the body in enough quantity to satisfy its demands. Fifteen other amino acids can be synthesized from C, O, H, N, S that we receive from food, and are considered nonessential in that sense. The eight essential amino acids in adults must be obtained intact from the foods we eat. This is one reason that vegetarian diets can be dangerous for those who do not know what they are doing. Vegetables are not normally considered a good source of high-quality protein. The quality of a protein depends on the relative proportion of essential amino acids it contains and the amount of such amino acids. Meat, fish, cheese, milk, and eggs all contain high-quality protein in fairly large amounts and therefore contain the essential amino acids. A food may be high in protein, that is, high in quantity but lack the correct proportion of essential amino acids and therefore be low in quality. Milk has a lower amount of protein than dry beans, but the protein of milk is of much higher quality. Individual vegetables do not generally contain all eight essential amino acids. Therefore, people on a strict vegetarian diet must eat mixtures of vegetables at the same time so that they may obtain all the essential amino acids. For amino acids to be used properly, all eight must be received in the stomach within approximately 4 hours of one another. If one or two essential amino acids are not present, the other six or seven will not be utilized as efficiently as if all were present. Therefore, those wishing to follow the vegetarian way should have some knowledge of the proper vegetable combinations to eat prior to beginning such a diet.

Protein Quality

There are many terms used to attempt to define the quality of a protein. The biological value (BV) is based on the nitrogen retained in the body in relation to the nitrogen absorbed by the body (remember all protein, but not carbohydrate and fat, contains nitrogen). Another commonly used term is the protein efficiency ratio (PER), which relates weight gain in rats to protein consumed with a reference protein and comparing weight gain. The PER of casein (milk protein) is usually taken as a standard at 2.5 and all other proteins are compared to this. For instance, if a test protein caused 80% of the weight gain of casein it would be given a PER of $80\% \times 2.5 = 2.0$. Another common term used with proteins is the net protein ratio (NPR). In conducting the NPR assay, one group of rats eats a test protein while a second group eats a diet that contains no protein. The NPR is defined as the weight gained by the test group plus the weight lost by the nonprotein group, divided by the protein consumed. A fourth term often used is net protein utilization (NPU), which is similar to NPR except that body nitrogen rather than body weight is used. NPU = (retained nitrogen ÷ food nitrogen) × 100. The NPU attempts to incorporate both BV and the digestibility of the food, since it relates food nitrogen rather than absorbed nitrogen, like the BV, to nitrogen retained.

Some nutritionists and food scientists feel that none of these terms adequately assesses both the quality and the quantity of food protein. They prefer another method of doing this, which they have named "biologically utilized protein" or "utilizable protein." This method supposedly recognizes the inherent interrelations between nutritional quality and quantity and applied nutrition. The term "biologically utilizable protein" may be defined as a food's crude protein content measured by chemical analyses and multiplied by the ratio of its NPR or NPU to the NPR or NPU of a standard protein such as casein.

There are other techniques, but it must be obvious by now that the degree of complexity is increasing. This is necessary for the scientist in the laboratory but not for the reader, so suffice it to say that there are methods available that provide an approximation of the quality of a protein based on absorption, growth, and digestability of the food.

Protein Metabolism

The importance of both adequate energy and nutrient intake is often magnified in developing countries where food intake may be marginal. The quantity of protein may be low because of low food intake and the quality may be poor since animal foods and by products are often lacking in the diet. Plant foods can easily provide adequate quantity and quality of protein if both enough food and the right variety are available. When this is not the case there may be problems with the protein deficiency disease, kwashiorkor. This disease, which is more prevalent in developing countries than in developed countries, normally occurs in children between two and five who are weaned from their mother's milk and begin to consume a diet of plant foods rather than milk. It is caused by a deficiency in the quality and quantity of dietary protein even though there is a normal intake of calories. (Marasmus, on the other hand, is a condition resulting from a caloric deficit that is usually also accompanied by a protein deficiency.) The many clinical symptoms of kwashiorkor include failure to grow, mental changes, accumulation of fluids in the tissues, changes in hair and skin, enlargement of the liver, and anemia. The ability of the afflicted child to combat infection is very low, and death is usually attributed to an infection such as measles or pneumonia that is not normally fatal in a healthy child.

Recently, many studies have indicated that the severe protein deficiency that exists in diseases such as kwashiorkor may be alleviated to a great extent by an increase in calories. If the number of calories from nonprotein sources is increased, then the protein ingested can be used to build and replace tissue rather than be used for energy. Adequate calories thus creates a "protein-sparing" effect. This concept has many ramifications for developing countries because it implies that the world food problem is due not to a protein shortage but to a calorie shortage, and we should be looking at more efficient methods of utilizing carbohydrate crops as food.

This "protein-sparing" effect may be easier to understand if we look at Fig. 1.2 where the breakdown and fate of dietary proteins in the body is shown. Dietary protein is broken down and absorbed in the body to form an amino acid pool. This pool provides amino acids to the body for three major functions: (1) energy production, (2) protein production, (3) nonprotein production. Keeping in mind that energy is one of the major requisites for life the body will satisfy this need from food first. If there is

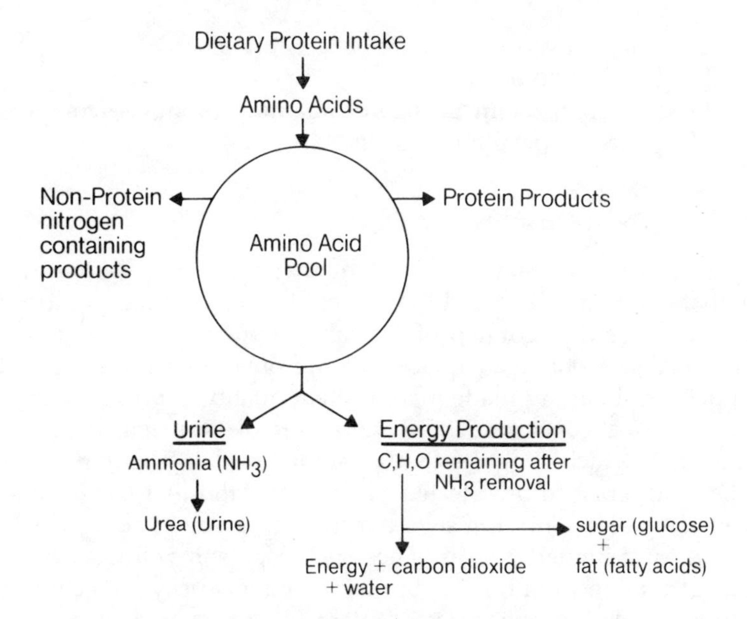

Fig. 1.2. The digestion of proteins and their metabolic fate.

sufficient food the body will use carbohydrates preferentially for energy and amino acids for protein and nonprotein production. However, if there is insufficient food the body will use amino acids for energy thus depleting the pool and leaving little for protein production. Therefore, increasing carbohydrates in the diet will allow all the protein present to be used for protein and nitrogen-containing nonprotein products. Thus "protein-sparing" occurs.

Proteins are of such importance in the body because of their many functions, which may be summarized as follows (Krause and Mahan 1979):

1. Repair worn out body tissues (muscle, skin, bone, nails, etc.). The build up is known as anabolism and the breakdown of tissue is known as catabolism.
2. Build new tissue (anabolism)
3. Provide sources of heat and energy
4. Produce body secretions and fluids (enzymes, some hormones, mucus, milk, and sperm, etc.)
5. Maintain osmotic balance (proper dilution and storage of fluids in the body)

6. Transport fat, iron, calcium, etc., in the blood
7. Produce antibodies
8. Produce nonprotein nitrogen-containing products (components of the genes, porphyrins in the bile, etc.)

Proteins in Food: Quantity and Quality

With all these functions it is obvious that there are potential risks in diets that eliminate all animal foods including eggs, milk, poultry, fish, and meat, since they contain high-quality protein.

It is possible to obtain adequate protein from only plant foods, but the potential for obtaining inadequate protein quality is present as well as other nutritional concerns. There are risks in diets such as the zen macrobiotic diet, which suggests that all animal proteins be removed from the diet and replaced by selected plant foods, and that water intake be limited. This diet is not too severe if tried by a young adult, say, a 20-year-old, for a limited period, since such a person will have had adequate nutrition most of his life, but it is extremely risky for a child. Such a regimen of eating only selected plant foods rather than milk and a wide variety of other foods could result in some irreversible metabolic changes in a child.

Foods that contain high quantities of high-quality protein are generally expensive. This is why kwashiorkor is seen most often in the developing nations of the world. One need only look at the prices of foods of very high biological value, such as milk, fish, and meat, to realize that these animal foods are not available even to some people in the United States as well as to many of the people in the developing countries of the world.

Certainly, a case can be made for replacing animal protein with vegetable protein, if the vegetables are chosen properly. Certain vegetables have a fairly well-balanced amino acid content and quite a high biological value. Soybeans, the bean family in general, and the nut family are good sources of protein, particularly in combination with certain other foods. Soybeans are slightly deficient in one of the essential amino acids, methionine, but this nutrient can be obtained from other food sources. In general, the grains, such as corn, wheat, and rice, are deficient in one or more amino acids and are called incomplete protein. Wheat, in particular, is deficient in lysine and threonine. This is a very important deficiency because wheat is the staple food for most of the world. The use of vegetables as a sole protein source, then, requires some knowledge of their protein quality. Many cultures have this knowledge and their diets reflect this. The American Indians did this when they consumed succotash, a mixture of corn and beans. Other examples are soybean curd and rice, rice and peas, and macaroni (pasta) and beans.

Another factor with vegetable protein is its digestibility. Vegetables are high in cellulose, a carbohydrate that is not digested by humans. Cellulose is bound intricately with protein in vegetables, and because of its nondigestibility it decreases the digestibility of the protein present—to about 70 or 80% at times.

In general, top-quality proteins are found in lean meat, poultry, fish, seafoods, eggs, milk, and cheese. The next best group of foods for protein are dry beans, peas, and nuts. Cereals, bread, vegetables, and fruits provide some proteins, but of lower quality. In general, plant and animal products that are related to reproduction and care of the young (e.g., eggs, milk) are excellent sources of protein, as you can see in the previous list, whereas leafy or stalky vegetables are poor sources.

Certainly, an important and increasingly popular development is the use of vegetable seed, leaf, fish, bacterial, fungal, and other alternative sources of protein that are flavored to taste much the same as meat. (This will be discussed in detail in Part II.) This can make a valuable contribution to nutrition (and one that will become even more valuable in the future), since the human diet should be about 12–15% of calories as protein and one-half of this should be of high quality.

It is interesting to note that the major sources of protein have changed in the United States since early in the century. This is not surprising, however, since in every society people begin to eat more animal products as they become more affluent. As a result, we now get about 68% of our protein from animal sources and 32% from plants, whereas we used to get protein equally from each. This also means that we now get more fat along with the animal protein and less carbohydrate with the decrease in plant consumption. One recommendation for health would be to return to the earlier practices and eat less animal food and increase our consumption of fruits, vegetables, and grains including bread, cereals, and pasta.

How much protein do foods contain? Table 1.2 will give us some idea of the protein in various classes of foods. The figures in Table 1.2 may vary, but you should be able to remember generally the approximate protein content of the food you choose.

Protein in Foods: Function

At this point one important thing should be emphasized about food—it must be eaten. Too often we get caught up in discussions of food components from only a nutritional view and forget that these same

Table 1.2. Protein Content of Some Foods

Food	% Protein
Leaf vegetables	Less than 1%
Stem vegetables	Less than 1%
Fresh deep green vegetables	2–3%
Mushrooms (fungus)	4%
Grains	10–15%
Fruit	0.5–2%
Beans, peas	5–8%
Nuts	20%
Breads	8–10%
Pasta (cooked)	5–10%
Cereals	5–20%
Hard cheese (cheddar)	30%
Soft cheese (cottage)	13%
Milk	3.5%
Butter	0%
Eggs	13%
Meat	25–30%
Poultry	25–30%
Fish	25–30%

components are responsible for taste, color, odor, texture, appearance, and overall quality—the factors upon which we base acceptance or rejection.

Proteins in food bind the fat to other protein in sausages and hot dogs, the coagulation of milk to make cheese depends on milk protein (casein) forming large insoluble curds when acted upon by acid, heat, or enzymes. Milk proteins also add flavor and stabilize foam in whipped cream. The protein in nonfat dry milk aids in baking and strengthening the dough. Egg protein helps coagulation, gel formation, emulsions, and thickening. Certain cereal proteins, like gluten in wheat, are elastic and expand to make bread. Seed proteins like soy help fat absorption and hold moisture.

It is apparent that proteins help us to have high-quality food as well as nutrients. In fact, one of the major reasons we often cannot substitute one protein for another in food is functionality. It is hoped, as time goes on, we will find ways of improving the functional qualities of alternative sources of proteins so that we can substitute them in foods to provide more food at a better price to more people.

Carbohydrates

Quick energy sources? Fattening foods? The cause of tooth decay? Empty calories? These are only a few of the terms that are often and mistakenly applied to carbohydrates. The truth is that carbohydrates should be our major source of energy and supply 50–60% of our total calories. Humans and animals do not store carbohydrates to any extent. They store fats and protein and preferentially use any carbohydrate present for energy or for conversion to fat or protein. Therefore, food from animals is generally high in protein and perhaps fat but low in carbohydrate. On the other hand, food from plants is high in carbohydrate and low in protein and fat. This is so because plants use carbohydrates for structural material (stalk, leaves, etc.) just as animals use protein (muscle) for structural material. This is a good generalization to remember if you are trying to increase or cut back on your consumption of fat, carbohydrate or protein. Plant foods such as vegetables and fruit provide us with nearly all our carbohydrate.

Carbohydrates are a group of chemical compounds made up of carbon (C), oxygen (O), and hydrogen (H). The individual carbohydrates in this group can vary in size and complexity. The simplest carbohydrate unit is known as a monosaccharide ("mono" means one or single and saccharide means sweet), that is, a single sweet unit. Common examples of this are glucose and fructose, which are the simple sugars in honey and fruit. When these two sugars join together, they lose two hydrogen and one oxygen (H_2O = water) to form a disaccharide called sucrose, which is white table sugar. Thus, the only difference between honey and table sugar is a molecule of water. These saccharides can continue to join together to form larger molecules just as amino acids join together to form large proteins. The building blocks of carbohydrates are monosaccharides as the building blocks of proteins are amino acids.

As they increase in size they form trisaccharides, then dextrins, then finally starch, glycogen, and the nonabsorbable carbohydrate fibers, cellulose, hemicellulose, gums, and pectin. Another fiber is known as lignin but it is not a carbohydrate.

The carbohydrates we normally consume that provide calories, are sugars and starches from plant foods. Animal foods do not contain appreciable quantities, with one exception. Milk contains lactose, a disaccharide known as milk sugar, and this is the only major dietary source of carbohydrates from animal food. Digesting lactose is a problem for certain people who are apparently deficient in the enzyme lactase, which is essential for the breakdown and absorption of lactose. Such "lactose-intolerant" people cannot digest milk properly; instead, they suffer from diarrhea and flatulence. This condition occurs in children as well as in adults. Obviously, milk cannot be considered a major food source for such people. They can consume some milk, however, and also other dairy products such as yoghurt and cheese, which are lower in lactose or more easily digested.

Included in the plant foods we ingest are fibers, but we cannot obtain energy from them since we do not have the necessary enzymes to break them down to monosaccharides so they may be absorbed. Grazing animals have bacteria in their intestinal tracts (actually in one of their stomachs) that can break down fiber and thereby make it available as an energy source. Fiber in plants acts in much the same manner as muscle in humans: it gives rigidity to the plant and acts as a stabilizing or binding force in its structure. In our discussion of proteins we mentioned that the proteins from plants are not as high in quality as those from animals. We noted that one of the reasons for this is that plant proteins are interwoven with fiber. Fiber does, however, have several important functions which will be discussed later.

As mentioned previously, the consumption of carbohydrates decreased in the United States from 1900 through 1982 (Fig. 1.3). However,

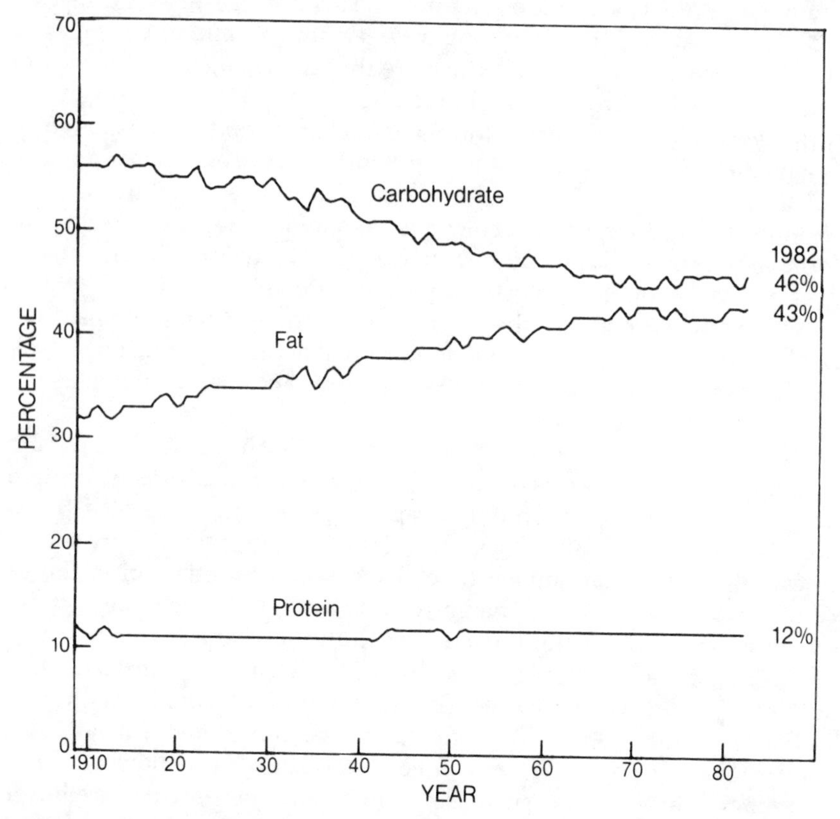

Fig. 1.3. Percent of calories from protein, fat, and carbohydrate in the US over the last 70 years. *From Wolf and Peterkin (1984).*

in the past few years there has been a renewed interest in fruits and vegetables and consumption is on the incrase. This is a positive step because health professionals recommend that 50–60% of calories be obtained from carbohydrate.

Digestion and Absorption

The absorption of carbohydrates from the intestines through the intestinal wall and into the bloodstream, like all other absorbed compounds, depends on their breakdown to simple units (monosaccharides), which are small enough to pass through the intestinal wall. In the mouth an enzyme called salivary amylase is excreted into the saliva and begins carbohydrate breakdown. It continues in the stomach where acid is released, followed by further breakdown in the intestines by both pancreatic and intestinal enzymes. The end product of nearly all digestible carbohydrate is the simple sugar glucose with some fructose and galactose, two other simple sugars. A schematic representation of carbohydrate absorption is shown in Fig. 1.4. The indigestible fibers pass through the body and exit in the feces.

The glucose, fructose, and galactose pass through the intestinal wall to the blood and from there to the liver. Some of the carbohydrate is stored as glycogen (the animal storage form of carbohydrate). The rest of the carbohydrate is converted to glucose and re-enters the blood-

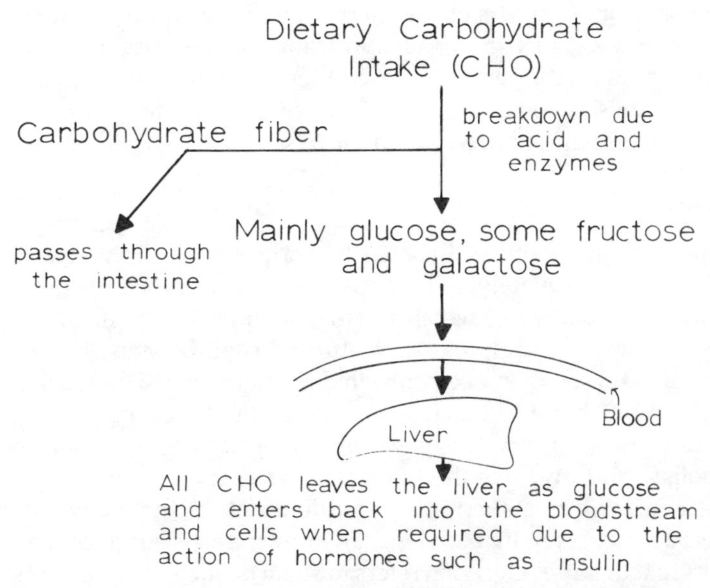

Fig. 1.4. The digestion of carbohydrates and their metabolic fate.

stream—remember all carbohydrate leaves the liver as glucose—to be used in one of several ways:

1. Production of energy in all cells and tissues
2. Storage as glycogen in muscle (small amounts)
3. Conversion to fat and stored as fat tissue
4. Conversion to other necessary carbohydrates
5. Supply C, H, and O for the manufacture of nonessential amino acids

Regulation of Blood Sugar

As is evident from the last section all digestible carbohydrate that we consume is finally converted to glucose, a simple sugar, and utilized by the body in that form. Glucose has many critical functions in the body. It provides energy, spares protein, is utilized in detoxification in the body, is necessary for normal fat metabolism, is the sole source of energy for the brain, and is a precursor for other essential compounds in the body. Obviously, with such important functions it is necessary that a constant supply of blood sugar be available. For this reason the body uses a host of hormones to regulate blood sugar levels.

Insulin is produced by the pancreas and causes the removal of glucose from the blood to be used in muscle cells, fat cells, and for the production of glycogen. This creates a decrease in blood sugar. Glucagon, another hormone produced by the pancreas, has an opposite effect in that it increases glucose production from the breakdown of glycogen, protein, and fat. Although there are other hormones involved in glucose utilization these two are the major ones responsible for glucose metabolism and the maintenance of blood levels.

Diabetes

In the disease known as diabetes the body is unable to metabolize and utilize sugars completely. There are two general types of diabetes: (1) juvenile-onset diabetes, which begins abruptly, early in life, and requires insulin to control, and (2) maturity-onset diabetes, which begins later in life and can often be controlled by diet alone. The exact cause of diabetes is not known, but it is a well-accepted fact that the intake of sugar does not cause it. It is generally thought to be an inborn error of metabolism and involves the hormonal systems of the body.

In general, when insulin is absent, deficient, or ineffective, it causes dangerous increases in blood sugar known as hyperglycemia (hyper means great or large, glycemia refers to sugar). Since sugar is not released from the blood, all those critical functions described previously, including energy production in the cells, do not occur. Excessive amounts of glucagon may also be released causing more sugar to empty into the blood, and there is little of the coordination of insulin and glucagon production that occurs in the normal state.

Excessive weight increases the risk of diabetes and is one more reason to control weight. Prior to the 1970s, diabetics were told to control all carbohydrate intake; this resulted in a high fat diet, which, along with the disease itself, increased the risk of heart disease. Current recommendations are to eat 50–60% of calories as carbohydrates, 12–15% as protein, and about 30% as fat, with simple sugars eliminated from the diet. As research continues we will learn more about this disease, its control, and dietary management. For instance, insulin pumps have been developed that continuously inject small amounts of insulin into the body, rather than the single or twice daily large injections currently used. Also, research has indicated that certain foods, including some sugar-containing foods may not be the villains in diabetes they were once thought to be, whereas other traditional foods may not be as good as they were once thought to be. However, all the results are not in yet, and the low sugar, high carbohydrate (50–60% of calories) diet should still be followed unless you are under the care of a qualified physician–dietetic team.

Another condition (the reverse of diabetes) results in low blood sugar. It is known as hypoglycemia (hypo means low, glycemia refers to sugar). This can be due to metabolic errors, which result in the overproduction of insulin, underproduction of glucagon or an imbalance in their relationship. It is rather rare but could cause fatigue and depression. The consumption of carbohydrates, including sugar, does not cause this disease. The disease is due to a hormonal imbalance, as described previously.

Energy Production

Glucose or blood sugar is the major fuel in the body for the production of energy. It is interesting to note that a survey among young gymnasts found that they thought energy was provided by sugar and vitamins only. Vitamins, as we will discuss later, do not provide any fuel for energy and all the macronutrients—carbohydrate, fat, and protein—are only able to provide fuel once they are converted to glucose. In some cases fat can be used directly as fuel for energy, but this is not the preferred metabolic route of the body since ketones are formed as end products. Ketosis is a condition that can result from the production of ketones in a low carbohydrate diet and can produce unpleasant and potentially dangerous side effects.

Muscles require fuel to move just as a car engine does. Cars use gas and muscles use food as fuel. However, the muscle can not use baked stuffed potatoes and a salad directly as fuel, it must first be converted to blood glucose which in turn is converted to a chemical known as adenosine triphosphate (ATP).

Adenosine is composed of two substances, adenine (a nitrogen-containing compound) and ribose (a sugar); this is in turn attached to three phosphate groups to make ATP.

$$\text{ATP} = \text{adenine} - \text{ribose} - 3\ \text{phosphate}$$

ATP is manufactured as needed. This manufacturing process requires energy to join the groups together, and it may be thought of as loading a spring. Thus, ATP can store energy just like a loaded spring and release it when the spring is triggered. When this happens the ATP releases energy by releasing one of its phosphate groups and forming adenosine diphosphate (ADP). The ADP can then be converted back to ATP with energy from food and thus the cycle can go on.

ATP can be produced in the body in two ways. One is anaerobically (meaning without oxygen) and the other aerobically (with oxygen). Anaerobic production occurs when quick energy is required like in a sprint or running up the stairs; but it is inefficient and can only provide energy for a short time. In this system, energy is produced initially by a compound in the muscle called creatine phosphate, which loses its phosphate to provide energy for the regeneration of ADP to ATP. There is only enough creatine phosphate to last for about 10 seconds of all out exertion, but it will get us up a flight of stairs. Following this, glucose provides energy anaerobically by a process that also produces lactic acid as an end product. The lactic acid builds up in the blood stream, causing inhibition of the process and fatigue so that the exercise can only last a few minutes. Obviously, this type of energy production is inadequate for long-term energy needs. Therefore, the body must have a much more efficient aerobic process to produce energy from glucose; it is a series of reactions known as the Krebs cycle. This process produces water and carbon dioxide, rather than lactic acid, as end products, which are easily excreted by the body and do not build up in the blood.

Nutritive Sweeteners

Any sweetener that provides calories may be defined as a nutritive sweetener. Thus, honey, molasses, white table sugar, corn sugars, and maple syrup are all nutritive sweeteners and are all carbohydrates. In fact, they owe their sweetening effect to the same chemicals—the two monosaccharides glucose and fructose, either alone or joined together to form the disaccharide sucrose.

Sugars and sugar products are generally refined even though the public tends to think of some as "natural." However, the term natural often only indicates less purity.

Corn sugars are made by extracting the starch from corn and then treating it with an enzyme to form various combinations of glucose and fructose. Maple syrup is made by boiling off the water and concentrating the sap collected from maple trees. It should be remembered that processing is done in part to concentrate, or rid the food of undesirable ingredients. For instance, raw sugar, which is obtained from sugar beets or sugar cane, plants as natural as the trees from which we get maple syrup, contains contaminants. This precludes the use of raw sugar by law in the United States, due to potential health hazards, without further refining. However, raw sugar is 99% sucrose, so that refining does not really change the sugar that much, aside from removing the potentially dangerous contaminants. Other sugars available in the marketplace are generally just a modification of those we have discussed. Turbinado sugar is produced when raw sugar crystals are separated in a centrifuge and then washed with water. It is off-white and its total sucrose content is usually 99%. Brown sugar is a mass of fine sugar crystals covered with a film of highly refined, colored molasses-flavored syrup. Brown sugars range in color from light to dark brown and are graded by number. Invert sugar is the result of heating a solution of sucrose in the presence of an acid or of treating the solution with enzymes. In this process, the sucrose breaks up into two monosaccharides—dextrose (D-glucose) and levulose (D-fructose)—which together form the disaccharide sucrose. The mixture of dextrose and levulose in equal weights is invert sugar. During the inversion, water combines chemically with sucrose to produce the two simpler sugars. Liquid sugars are clear solutions of highly purified sugar and water. Liquid sugars are available in uninverted, partially inverted, and completely inverted grades.

The nutritional value of sugar and other nutritional sweeteners seems to be misunderstood by the consumer. This is unfortunate, because these simple chemical compounds play an important role in our food as well as supplying energy.

Their principal use in food is to sweeten it, making it more palatable as well as satisfying the natural desire for sweet tasting foods. It is interesting to note that an unborn fetus will make sucking movements only if a sweet solution is injected into the amniotic fluids of the mother. Other flavors do not produce this response, indicating that this desire is inborn. It seems more than coincidental that in nature most toxic plants are bitter while those that are safe are sweet. The use of sweeteners allows us to eat many foods that we might not eat if they were not sweetened. This is exactly the reason that we like ripe fruit. As it ripens more sugar is formed, increasing its desirability.

Sugars, in fairly high concentrations, also act as preservatives in such foods as jams, jellies, honey, syrups, and candies. They "tie-up" or bind water, making it unavailable for bacterial growth and thus preventing spoilage. They serve as food for yeast in making bread, pickles, and alcohol and as food for bacteria at lower concentrations in making yoghurt. They add to the crust color and flavor in baked goods, to the color of baked beans, and hold moisture in some food products so that they remain palatable longer. They provide bulk and texture in food as well as controlling the freezing point and crystallization in ice cream.

In terms of potential risk, there is very little from the moderate use of sugars in a varied diet. This does not mean that you should eat six candy bars a day, but it does mean that sugar use in moderation to increase palatability and keeping qualities, as well as for providing energy, has a role in a healthy diet. To illustrate such use one need only examine the use of sugar in cereals. A good breakfast should supply approximately one-fourth of the daily requirements for energy, protein, and most vitamins and minerals. This requirement may be met for a child as follows:

1 cup presweetened fortified cereal with milk
2 slices toast with butter or magarine
4 oz of orange juice
8 oz of whole milk (fortified with vitamin D)

In this breakfast the presweetened cereal provides a good flavor that most children accept. It also supplies about 3% of the child's total daily sugar intake, and one-fourth of the U.S. Recommended Daily Allowance of vitamins A and C, thiamin, riboflavin, niacin, vitamins B_6 and B_{12}, and iron.

The amount of sugar in presweetened cereal, which many are concerned about, is equivalent to the sugar in one tablespoon of jelly, one tablespoon of pancake syrup, or two canned peach halves served with two tablespoons of the fruit syrup. This is not an inordinate amount of sugar in a *well-balanced meal*. Of course, a good breakfast does not have to contain a cereal, but with today's lifestyle cereal seems to fulfill a consumer demand. This is particularly true with children and with the elderly. In both cases, studies have shown that ready-to-eat cereals provide an important nutritional component to the diet. Remember that there are no junk foods, or super foods, only junk diets or good diets. It is the combination of foods that we eat on a daily basis that provides us with the nutrients, not the ingestion of only one type of food.

Sugars do not cause heart disease, hypoglycemia, diabetes, or obesity as indicated by National Scientific Committees. Once again, however, it should be stressed that, like all foods, they should be part of a balanced diet and used in moderation.

Americans consume about 10% of their total calories as sugar and 7% from other sweeteners, as shown in Fig. 1.5. However, these are "disappearance figures," which means they represent what is purchased, not

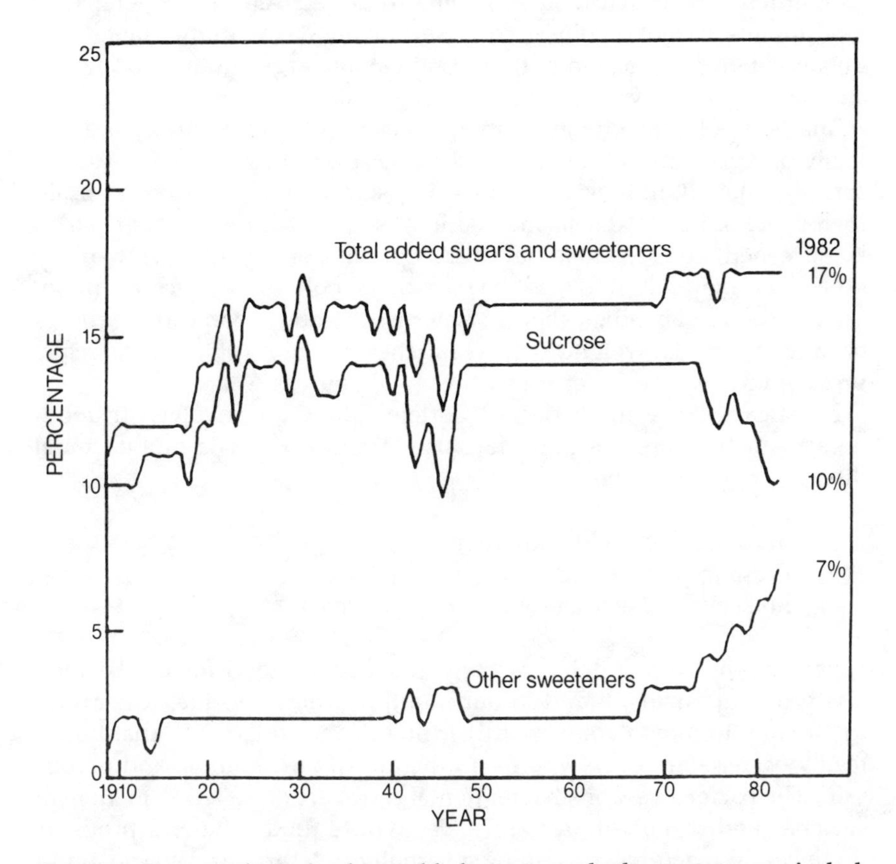

Fig. 1.5. Percent of calories from added sucrose and other sweeteners, including dextrose, cane, corn, maple, and refiners syrups, sorgo, molasses, and honey over the last 70 years. *From Wolf and Peterkin (1984).*

what is eaten, and do not include waste. Therefore, the amount of sweeteners actually eaten is considerably lower. These levels certainly seem to fall within the moderate range that was discussed previously; those that eat more should cut back. Remember, to maintain your correct weight range, reduce total calories, decrease your consumption of meat, and increase your consumption of fresh fruit, grains, cereals, bread, pasta, and vegetables.

Are carbohydrates fattening? Certainly they are fattening, in that they contain 4 calories per gram, the same as protein. However, carbohydrates and proteins each contain less than one-half the number of calories per gram that fat contains. Relatively speaking, then, carbohydrates are not particularly fattening. A medium-size potato contains only about 75 calories. Nevertheless, it has come under scrutiny by dieters, who tend to view it as fattening and push it away at the dinner table. The potato is not the problem here, but rather the large amounts of butter, sour cream or gravy that are placed on it.

The body obtains energy in many ways, and carbohydrates are certainly one very important source of energy. Carbohydrates are also essential to health: in their absence—for instance, when one is on certain high-protein diets—ketosis may occur. As far as the quick-energy claim is concerned, one food material will not provide energy quicker than any other food material. It is true that simple carbohydrates such as monosaccharides are absorbed slightly faster by the body than carbohydrates of larger molecular weight, such as starches, but none of these will really provide energy any faster than other food sources.

Do sugars cause tooth decay? Certainly they are involved in tooth decay, but they are not the sole cause. Tooth decay is a multifactorial disease dependent upon

1. Individual susceptibility (genetic)
2. Presence of a fermentable carbohydrate
3. Presence of bacteria (*Streptococcus mutans*)

Sugar or other fermentable carbohydrates act as food for the bacteria that grow on our teeth and in our mouth, forming plaque. It is rather frightening to most people that their mouths are full of bacteria, but in reality bacteria are everywhere—in the air, in our mouths, and on our skin. The bacteria responsible for dental caries are so-called fermentative bacteria, and carbohydrates are their favorite food. The end products produced by bacteria consuming a carbohydrate are mainly acids, and these dissolve the tooth enamel resulting in decay.

The longer that bacteria and food stay in our mouths, the more acid will be formed and the greater the decay. It is not the quantity of sugar we consume as much as it is the frequency with which we consume it, that is, how often we eat it during the day. The best advice is not to let a child have a continual bottle with juice (sugar and acid) or sweet drinks in it, and do not sip or eat sweets continuously. Rinse your mouth, brush your teeth, floss, and drink fluoridated water. Fluoridation of water has reduced caries in children by at least 50% in 5 years in every area where it has been used. It makes the teeth more resistant to acid, thereby protecting them from decay. The enamel of the teeth contains a substance called hydroxyapatite, which is made of calcium and phosphorus in the form, $Ca_{10}(PO_4)_6(OH)_2$. When we ingest fluoride, this material becomes fluorapatite as follows: $Ca_{10}(PO_4)_6(OH)_2 + 2F \rightarrow Ca_{10}(PO_4)_6F_2 + 2(OH)$. Fluorapatite is more stable, more acid resistant, and about 100 times less acid-soluble than hydroxyapatite and therefore does not allow decay to occur as easily.

Honey presents some risks in infants under one year of age. Since it is a natural product, we have less control over what it contains, and some honey is contaminated with spores of the bacteria *Clostridium botulinum.* This bacteria, along with its spores, is generally harmless if ingested because it must grow and produce a toxin in food as explained in Chapter 4. However, in infants under one year of age the spores can grow into cells and produce a toxin inside the body; this cannot occur in children over one year of age or in adults. The toxin thus produced can cause death; and there have been a number of infant deaths attributed to botulism from honey.

Aspartame, a very low calorie sweetener sold under the trade name "NutraSweet" is not a carbohydrate but a combination of two amino acids. Technically, it is a nutritive sweetener but for practical purposes (very low in calories) it can be classified as a nonnutritive sweetener like saccharin.

Fiber

Our grandparents used to call it "roughage," and knew that it was a part of the diet that helped to keep one "regular." Today we know it as a group of compounds called fiber, and we are learning more about their physiological effects every day.

Much of our current interest in fiber is due to the observations of Dr. D. P. Burkitt, a British physician who noticed a correlation between low fiber and "western type diseases." This was not a cause and effect relationship but nonetheless was convincing enough to spur further research.

Dietary fibers may be defined as those components of a food that are not broken down by enzymes and acids in the human and therefore are to large to be absorbed through the intestinal walls into the bloodstream. They include hemicellulose, cellulose, lignin, pectin, gums, and mucilages, as well as other minor carbohydrate components. Crude fiber—a term often erroneously used for dietary fiber—is, on the other hand, the components of a food remaining after treatment in the laboratory with hot acid and hot alkali. This is a much more severe treatment than food is given in the body and the only compounds to survive it are cellulose and lignin.

With all the press coverage, advertising, and publicity given to fiber, there is understandable confusion as to both its benefits and potential risks. It has been said to cure or prevent a variety of disorders and diseases including

Constipation	Diabetes
Diverticulosis	Appendicitis
High cholesterol	Obesity
High blood lipids	Gallstones
Hemorrhoids	Phlebitis
Varicose veins	Dental decay
Heart disease	Irritable bowel
Colon-rectal cancer	Ulcerative colitis

Obviously, all of these claims cannot be completely true, or we could do away with nearly all diseases by simply eating fiber. However, there is certainly evidence available that fiber plays some role in some of these diseases and often can alleviate symptoms.

Constipation is clearly helped by the ingestion of fiber. In general, cellulose and hemicellulose, are the fibers that most often aid constipation, and these are found in abundance in whole grains such as whole wheat, barley, rice, etc., or in cereals and in many unpeeled vegetables. These fibers tend to keep the stool soft and allow easy passage through the intestine. For this reason they are also thought to be of possible help in the treatment of disorders that might arise due to straining during a bowel movement, which creates undue pressure on the intestine. Just as a plugged garden hose can rupture from too much pressure when water expands in it after being left in the sun, the intestine might give out in certain places, resulting in such diseases as diverticulosis, appendicitis, hemorrhoids, etc. There is no definite proof that fiber will cure these conditions, but evidence suggests that it might help.

The water-soluble fibers, pectins, gums and mucilages, which come mainly from plant cells and fruits, seem to play a role in lowering blood lipids (fats) such as cholesterol. One of the reasons for this action is thought to be that the fiber binds bile acids and cholesterol in the intestine and causes them to be excreted in the feces. Normally, cholesterol in the liver produces bile acids, which are excreted into the intestine. These bile acids are then reabsorbed back to the liver and recycled. This is known as the "enterohepatic cycle." The fiber interrupts this cycle causing more cholesterol to be used to make up for the bile acids not returned to the liver. Thus the lipid level in the body is lowered.

Several studies have shown a hypoglycemic (lowering of blood sugar) effect of certain fibers, particularly those from the legume or bean family of vegetables. This effect has allowed some insulin-dependent diabetics on a very high fiber diet to stop the use of insulin. The type of diet used may have some other nutritional risks, however, and should not be considered as a replacement for insulin unless done under the care of a physician.

There has not been any cause and effect relationship shown between cancer and dietary fiber. Some epidemiological studies indicate a correlation between a low incidence of colon cancer and high fiber diets, but some of the same studies indicate a low incidence of stomach cancer in countries with lower fiber diets such as the United States. The beneficial action of fiber in such disorders has been attributed to the possibility that a high fiber diet will alter and inhibit certain microorganisms in the intestine that might produce carcinogens and to the fact that fiber speeds up the passage of all material, including carcinogens, through the body.

The role of dietary fiber in weight loss has not yet been proven. The reasons why fiber may contribute to weight loss are obvious since fiber adds bulk to the diet without calories. It has also been theorized that the faster passage through the body does not allow as much absorption of macronutrients and therefore fewer calories are contributed. However, it is safe to say that, if fiber helps, it is only one more link in the complex chain of events required to lose and then maintain weight loss.

In addition to the obvious beneficial effects to the intake of fiber, there are risks as well. The binding action and the increased rate of passage of food not only removes bile acids but also removes some nutrients such as minerals. Therefore, our desire for a "quick fiber fix" to good health should be tempered by the fact that abuse and overuse of fiber in our diet has risks. The best way to increase fiber in the diet is to follow the advice you will hear again and again in this book: reduce intake of animal foods and increase your consumption of grains, fruits, vegetables, cereals, pasta, and bread.

Remember that an increased fiber diet means a diet high in carbohydrates and therefore low in fat. Such a diet will have the potential for good health not only through fiber but due to low fat, high carbohydrate and low calories. No one component will give us good health; food gives us good health!

Carbohydrates in Food

The major food supply for the world is carbohydrates. Many in the western world do not realize this fact, but indeed it is true. In some parts of the world the carbohydrate source is wheat or rice, in others it is corn and beans, manioc, sweet potatoes, and potatoes.

Carbohydrates are found mainly in plants. The exceptions are small reserves of glycogen in animal tissue and lactose (milk sugar) as discussed previously. In plants carbohydrates act as structural material, as food reserves, and to bind water. The carbohydrates that provide energy in the diet are starch and sugars, which, like fibers, have other potential benefits.

In general, vegetables are 95% water, 3–5% carbohydrate, with small amounts of protein and minerals. Since vegetables are not sweet it may be concluded that they are low in sugar and therefore contain mostly starch, fiber, and water.

Fruits are 85–95% water, and if sweet are high in sugar, containing 10–15% sugar, some starch, fiber, protein, and minerals.

Legumes, peas, and beans are lower in water and higher in carbohydrate, containing 16–20% starch and fiber. Grains, pasta, and potatoes are 20–30% starch with some fiber. Dry cereals are about 80% carbohydrate, containing sugar and fiber. Flour, cookies, and crackers are 70–75% carbohydrate, bread 50% and dried fruits, preserves, and jams 70%—the latter mainly sugar. It is interesting to note that jams are often called "preserves." This is because fruit and sugar are boiled to evaporate the water, leaving a high concentration of sugar which "preserves" the fruit by binding the water and making it unavailable to microorganisms.

All carbohydrates have very important functional properties as well as nutritive properties in food. Sugars preserve food and keep it moist by binding water, as in jams and semi-moist food. They give body, texture, and color to food as well as contributing sweetness. Larger carbohydrates, like starches and fibers, act as thickeners, fillers, and as the structural base for foods such as cakes, rolls and breads.

It is interesting that we produce white flour for functional reasons as well as reasons of acceptability, although it is becoming more acceptable to eat whole grain baked goods such as bread. Flour made from wheat contains appropriate amounts of the protein gluten, or wheat protein, which allows wheat to be structured into forms such as bread and cakes. Certain other grains (such as barley) do not contain gluten and for this reason cannot be used in making these commodities. (Barley can be fermented to make liquors, such as whiskey and beer.) One of the major reasons for milling flour, of course, is that people had been accustomed to eating white flour. In the Orient white rice is a status symbol, and many people do not want to eat brown rice since it indicates a lack of status. But there are functional reasons for milling flour. In the milling process, the large amount of fat in wheat grain is removed. The removal of fat permits the flour to be stored for longer periods of time. For household use, whole-wheat (unmilled) flour may be suitably stored in the refrigerator. However, if we consider the millions of tons of flour used worldwide, the matter of storage becomes a problem. In the humid climate that exists in many parts of the world, the fat would become rancid making the flour inedible as well as less nutritious.

The overall benefits of increased use of carbohydrates in the diet on an individual, national, and worldwide basis are clear. Carbohydrates are a relatively inexpensive crop with a "protein-sparing" effect that have major nutritional attributes as well as critical functional properties in food.

Alcohol

Alcoholic beverages, unfortunately, constitute another major caloric source for a large segment of society. Alcohol is produced by the fermentation of a carbohydrate compound by a yeast. The metabolism of alcohol by the body requires a rather large amount of water. One ounce of whiskey requires about 4 oz of water in order to be metabolized. Most people who drink have at some point undergone the "thirst syndrome" the morning after. This occurs when not enough other liquids are consumed along with the alcohol. Therefore, when it is extremely hot it is better to have a couple of glasses of water rather than several strong drinks, because the body has already lost water and will lose even more as the alcohol is metabolized.

Moderate use of alcohol by an otherwise healthy adult does not seem to pose major risks. However, moderate use implies no more than two single strength drinks per day. The per capita consumption of alcohol in the United States is 76 calories per day. Since many people do not drink, that number is higher for those who do.

The use of any alcohol by a pregnant woman poses the risk of "fetal alcohol syndrome," which has been linked to birth defects, behavioral problems, and more recently to the incidence of cancer in the offspring.

Immoderate use of alcohol by anyone may lead to ill health, alcoholism, and death due to physical debilitation or accident. A healthy lifestyle does not have room for more than about 100 calories per day from alcohol.

Fats

Fats—or, more correctly, lipids—are chemical compounds containing carbon, hydrogen, and oxygen. They differ from carbohydrates in that they contain a much lower ratio of oxygen to carbon and hydrogen and as a result provide more than twice the calories of either protein or fats. Fats produce 9 calories per gram as compared to 4 calories per gram for protein and carbohydrates.

The major building blocks of fats are the fatty acids. These compounds can have 4 to 30 carbon atoms linked together in what is called a "chain" with hydrogen and oxygen atoms joined to the side of the chain as shown in Fig. 1.6. These fatty acids may be joined to a glycerol molecule which can accept 1 (mono), 2 (di), or 3 (tri) fatty acids forming what are called monoglycerides, diglycerides, or triglycerides (Fig. 1.6).

Most natural fats are triglycerides, but there are some mono- and diglycerides, free fatty acids, and fatty acids combined with other materials. All fats are lipids but not all lipids are fats since the lipids also include such compounds as cholesterol, which is a sterol and does not have the fatty acid configuration.

The term lipid includes a wide group of compounds which are insoluble in water. Lipids are classified as follows:

Simple lipids:
1. Fatty acids
2. Mono-, di-, triglycerides
3. Waxes

Compound lipids (Combinations of fatty acids with compounds other than glycerol):
1. Phospholipids (phosphoric acid + fatty acid)
2. Phosphoglycerides (phosphoric acid + glycerol)
3. Glycolipids (carbohydrate + fatty acid)
4. Lipoproteins (proteins + fatty acids)

Derived lipids:
1. Sterols, e.g., cholesterol
2. Fat-soluble vitamins

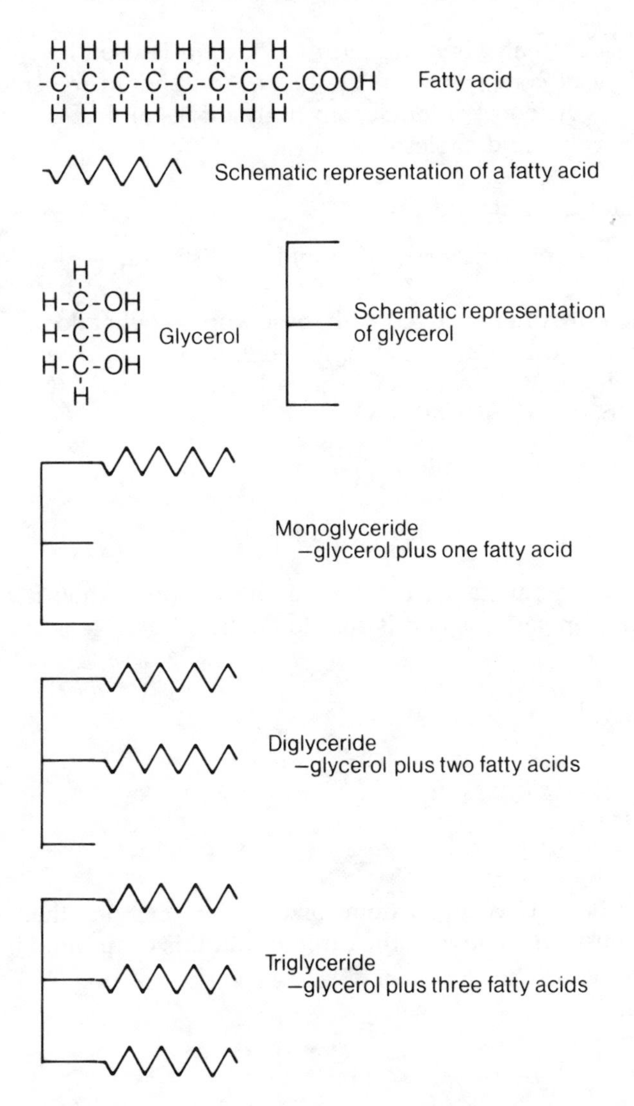

Fig. 1.6. The chemical make-up of commonly eaten fats.

It can be seen that when the term fat or lipid is used it can cover a whole variety of compounds in the body. You should remember that nearly all fats in food are long-chain triglycerides and of course some foods contain the lipid cholesterol.

Saturated and Unsaturated Fats

There is one other point that should be discussed—saturated and unsaturated fats.

Consumers are often deluged with these words—what do they mean? The degree of saturation of a fat simply refers to the number of hydrogen atoms on the fatty acid chain. The carbon atoms in a fatty acid have four points of connection or "bonds."

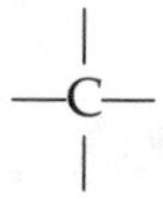

```
  |
—C—
  |
```

If each carbon atom has two hydrogen atoms on it then it has one "bond" between carbons and is said to be "saturated."

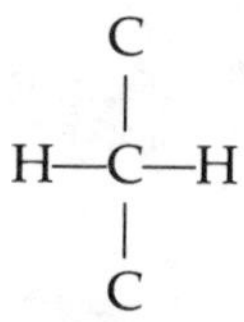

```
   C
   |
H—C—H
   |
   C
```

If there are not two hydrogen atoms on each carbon atom, then a "double bond" must exist between the carbons and this compound is said to be "unsaturated."

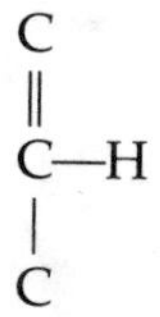

```
C
‖
C—H
|
C
```

Therefore, the terms really mean saturated with hydrogen or not saturated (unsaturated) with hydrogen. This is also illustrated in Fig. 1.7 where the term hydrogenation is also introduced to indicate the addition of hydrogen to an unsaturated fat to cause it to become saturated. An unsaturated fat therefore has fatty acids which are unsaturated. If the

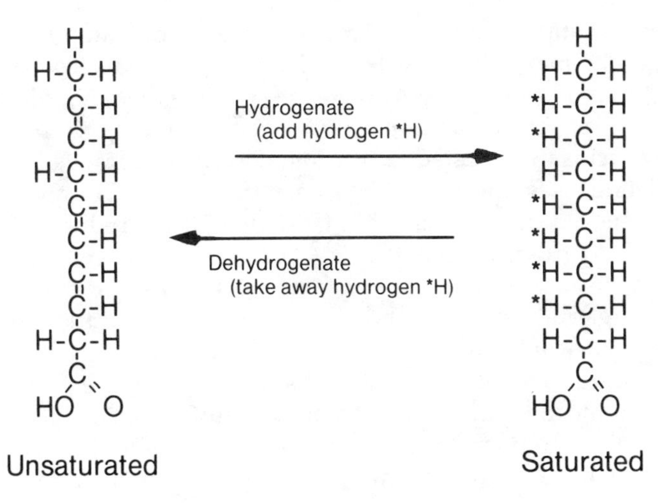

Fig. 1.7. The hydrogenation of unsaturated to saturated fats.

fatty acid has two hydrogen atoms less and therefore contains one double bond it is said to be monounsaturated (mono = one). If it contains two double bonds it is diunsaturated (di = two) and if it contains many double bonds it is said to be polyunsaturated (poly = many).

The differences due to saturation and unsaturation in a fat, although rather simple chemically, have great significance in physical, physiological, and metabolic terms.

In general, saturated fats are solid at room temperature and usually come from animal sources. Therefore, meats, chicken, butter, ice cream, whole milk, yogurt and cheese contain more saturated than unsaturated fats. Exceptions to this rule are the two vegetable oils coconut oil and palm oil. Monounsaturated fats are often liquid at room temperature and are found in olive oil and peanut oil, for example.

The polyunsaturated fats we hear so much about are in actuality oils, since they are liquid at room temperature (an oil is simply an unsaturated fat which is liquid at room temperature). They are genearlly of vegetable origin and are found in corn, soybean, safflower, and sesame seed oils. The exception is fish, which contains polyunsaturated fats.

It is possible to change not only the type of saturation in a fat, which will change its physical properties, but also the type of fat in pork, beef, and poultry by changing the type of feed. However, because the flavor and mouthfeel of a food is closely related to the type of fat it contains, we must be careful not to modify the food so much that we make it inedible.

A major example of fat modification is the production of margarine. By definition, margarine is a butter-like spread made from vegetable oils. Something has to be done to the oils to make them hard like butter. This is accomplished by adding hydrogen and changing the polyunsaturated fatty acids to saturated fatty acids. Look at a package of margarine you will often see the term "hydrogenated oils." The question that should be asked at this point is "How does this product differ from animal fats if you saturate all the fatty acids?" The answer is that not all the fatty acids are saturated but just enough to make the product solid. Not all margarines are alike, however, and you should know the difference if you want to obtain the most polyunsaturated fatty acids for potential health benefits. To do this you must read the ingredient label (Fig. 1.8). Some margarines will contain coconut and palm oils, which, if you remember, contain saturated fatty acids. You should look for margarines that have a polyunsaturated oil as their first ingredient. Un-

NUTRITION INFORMATION
(PER SERVING)

SERVING SIZE	1 TBSP (14g)
SERVINGS PER CONTAINER	32
CALORIES	100
PROTEIN	0g
CARBOHYDRATE	0g
FAT	11g
CHOLESTEROL (0 mg per 100 g)	0mg
SODIUM	110mg
PERCENTAGE OF U.S. RECOMMENDED DAILY ALLOWANCES (U.S. RDA)	
VITAMIN A	10%

CONTAINS LESS THAN 2% U.S. RDA OF PROTEIN, VITAMIN C, THIAMINE, RIBOFLAVIN, NIACIN, CALCIUM, IRON.

INFORMATION ON CHOLESTEROL CONTENT IS PROVIDED FOR INDIVIDUALS WHO, ON THE ADVICE OF A PHYSICIAN, ARE MODIFYING THEIR TOTAL DIETARY INTAKE OF CHOLESTEROL.

INGREDIENTS: VEGETABLE OIL BLEND (PARTIALLY HYDROGENATED SOYBEAN OIL, LIQUID CORN OIL AND PARTIALLY HYDROGENATED CORN OIL,) WATER, SALT, WHEY, VEGETABLE LECITHIN, VEGETABLE MONO AND DIGLYCERIDES. SODIUM BENZOATE AND CITRIC ACID ADDED AS PRESERVATIVES. ARTIFICIALLY FLAVORED. COLORED WITH BETA CAROTENE. VITAMIN A PALMITATE ADDED

SERVING SIZE (1 TBSP.)	14g
SERVINGS PER CONTAINER	32
CALORIES	100
PROTEIN	0g
CARBOHYDRATE	0g
FAT (100% OF CALORIES)	11g
POLYUNSATURATES*	3g
SATURATES*	2g
CHOLESTEROL* (mg/100g)	0mg
SODIUM	115mg

PERCENT U.S. RECOMMENDED DAILY ALLOWANCE (U.S. RDA)
VITAMIN A 10%
CONTAINS LESS THAN 2% OF THE U.S. RDA OF PROTEIN, VITAMIN C, THIAMINE, RIBOFLAVIN, NIACIN, CALCIUM AND IRON.
*THIS INFORMATION ON FAT AND CHOLESTEROL CONTENT IS PROVIDED FOR INDIVIDUALS WHO, ON THE ADVICE OF A PHYSICIAN, ARE MODIFYING THEIR TOTAL DIETARY INTAKE OF FAT AND CHOLESTEROL.
PREPARED FROM LIQUID CORN OIL, PARTIALLY HYDROGENATED CORN OIL, PASTEURIZED SKIM MILK AND/OR WATER AND NON-FAT DRY MILK SOLIDS, SALT, LECITHIN, **ARTIFICIAL FLAVOR ADDED, ARTIFICIALLY COLORED WITH BETA CAROTENE,** VITAMIN A PALMITATE ADDED.

Fig. 1.8. Examples of margarine labels showing ingredients including saturated and unsaturated oils.

hydrogenated liquid oils are the major ingredient in these margarines, and the manufacturer adds just enough partially hydrogenated oil to make them solid. This mixture of polyunsaturated oil and partially hydrogenated oil is then emulsified (making a stable mixture of two generally insoluble components) with water, at a level of 80% oil and 20% water, which is the composition of both butter and margarine.

The Role of Fat in the Diet

One hears so much about the risks of fat in the diet that the fact that fats are essential nutrients is often forgotten. Fats serve as a concentrated form of energy for the body and as such are used by the body to store surplus energy for times of need. If the body used carbohydrate or protein as storage, we might be twice the size we are now because we would have to store twice as much carbohydrate and protein for the same amount of energy from fat. They provide 4 calories per gram and fat provides 9 calories per gram.

In addition to energy storage, fats cushion the vital organs and act as an insulating layer for the body thus helping to maintain body temperature. They act as carriers of the fat-soluble vitamins and aid in their absorption, as well as providing us with flavor in food and a feeling of fullness.

We also require three polyunsaturated fatty acids, linoleic, linolenic, and arachidonic, for the maintenance of health and life. As well as being a part of many other vital lipids these fatty acids are precursors of a group of compounds called the prostaglandins, which act in the regulation of blood pressure, heart rate, lipid formation, and the central nervous system.

The body can synthesize linolenic and arachidonic from linoleic acid, so it is the only one we really need and is called an essential fatty acid. It is in plentiful supply from most vegetable oils, such as corn oil and soya oil, but not from peanut and olive oils. Occasionally, a skin condition is noted in infants on a very low fat diet due to a lack of essential fatty acids, but this is very rare in adults.

Even cholesterol has vital functions in the body; in fact, the body manufacturers about 70% of its own cholesterol to make sure there is enough. Cholesterol is an essential part of the production of other important steroids like the bile acids, the sex hormones, and the adrenocortical hormones. It is converted to 7-dehydrocholesterol in the body, which finally forms vitamin D, and has several other important functions. Therefore, it is obvious that lipids play an important role in our well being but like other nutrients should not be overused.

Digestion and Absorption

The digestion and absorption of fats depends somewhat on the chain length of the fatty acids in the fat. Some digestion of short chain fatty acids begins in the stomach by the action of the enzyme gastric lipase. Interestingly the presence of fat in the diet lengthens the time food is retained in the stomach. The fat must be made more soluble in water to be fully digested and absorbed. This is accomplished in the intestine by bile, which is manufactured from cholesterol in the liver and secreted into the duodenum. The bile emulsifies the fat, which allows it to be attacked and broken down by other lipases (enzymes) in the intestine. The smaller particles are also emulsified by bile and carried into the intestinal wall where the fat portion is finally coated with protein to form a lipoprotein called a chylomicron. Chylomicra pass into the lymphatic system and enter the bloodstream where they are carried to the liver. The bile is recycled back to the liver and reused in a continual cycle known as the "enterohepatic cycle." Remember, fiber is thought to lower blood cholesterol by carrying bile out of the body thus breaking this cycle and making the liver use blood cholesterol to manufacture more bile.

About 70% of the fats are absorbed via the lymph system, but the remaining 30% consist of short-chain fatty acids, which are more water soluble and can be absorbed directly into the bloodstream to the liver.

Fat in the U.S. Diet

The per capita consumption of fat in the American diet is too high. It is estimated that Americans now eat about 46% of their calories as fat (Fig. 1.3); this number should be reduced to about 30%. As explained previously, this has happened because Americans, like all people who become more affluent, began to obtain more protein from animal foods and less from plant foods. In 1900, the ratio of plant to animal protein in our diet was 50:50. Today it is about 32:68. This means that along with the animal protein we get more fat and less carbohydrate than we would get if we were eating more plant protein.

Table 1.3 shows a general categorization of the fat content of various foods. Obviously, this is a list of desirable foods both from the pleasure and the nutrients they provide. Some foods such as butter, margarine, and vegetable oils listed in Table 1.3 contain nearly 100% fat. This is called "visible" fat. Other foods like meat, contain some "visible" fat, which can be trimmed off and some "invisible" or hidden fat which cannot be removed. Interestingly, this hidden fat is what makes the meat so juicy and flavorful. Still other foods such as egg yolk, nuts, ice cream, pastries, breads, and cheese contain all "invisible" fat.

Table 1.3. The Fat Content of Various Foods[a]

Fat (%)	Food
90–100	Salad and cooking oils and fats, lard
80–90	Butter, margarine
70–80	Mayonnaise, pecans, macadamia nuts
50–70	Walnuts, dried unsweetened coconut meat, almonds, bacon, baking chocolate
30–50	Broiled choice T-bone and porterhouse steaks, spareribs, broiled pork chop, goose, cheddar and cream cheeses, potato chips, french dressing, chocolate candy, butter cream icing.
20–30	Choice beef pot roast, broiled choice lamb chop, frankfurters, ground beef, chocolate chip cookies
10–20	Broiled choice round steak, broiled veal chop, roast turkey, eggs, avocado, olives, chocolate cake with icing, french fried potatoes, ice cream, apple pie
1–10	Pork and beans, broiled cod, halibut, haddock, and many other fish, broiled chicken, crabmeat, cottage cheese, beef liver, milk, creamed soups, sherbert, most breakfast cereals
Less than 1	Baked potato, most vegetables and fruits, egg whites, chicken consomme.

[a] From Leveille and Dean (1978)

How then do we cut down on our fat intake? We can moderate our intake of those foods that contain fat. From a nutritional point of view it would be most unwise to eliminate meat and dairy products from the diet for many reasons. For instance, without dairy products, milk, cheese, yoghurt, etc., it is difficult to obtain enough calcium. This could lead to osteoporosis, particularly in women, later in life and the high risk of spontaneous fracture of the bone. Meats provide us with readily available forms of iron and zinc, two minerals difficult to obtain elsewhere. We might also decide to eat more of the "visible" fats in the form of polyunsaturated fats by using more products that contain vegetable oils to add to our foods or to cook in.

Fats and oils, like the other macronutrients, have important functions in food. They contribute to flavor, juiciness, and palatability of food. They are used to cook food and also contribute to color, flavor, and texture as in crisp, golden brown french fries. They contribute "shortness" to baked goods when added as shortening and contribute greatly to the enjoyment of salads when used as salad dressing, oils, or mayonnaise. Many consumers do not realize that mayonnaise in foods also helps to retard bacterial growth and thus helps to prevent food poisoning because of the vinegar it contains.

Fats are important emulsifiers in some food and also act as anti-staling agents in others.

Obviously, we would have difficulty in doing without fats in foods. Once again, for nutritional reasons, we should simply reduce the size of the large cuts of meat we eat and increase our consumption of fresh fruits, vegetables, grains, cereals, pasta, and breads.

Health Risks of High Fat Diets

Although the death rate in the United States has dropped to an all time low, and life expectancy has reached a new high, heart disease still remains the No. 1 killer with cancer No. 2 and stroke No. 3.

There is no single cause of heart disease, and several major risk factors—like heredity—you cannot control. Others, however, like obesity, high blood pressure, smoking, diet, lack of exercise, and stress, are controllable. This also can be frustrating and confusing when "good old Uncle Joe," who smoke, drank, gorged himself, and never lifted a finger, lived to be 100 years old, while Uncle Charlie, who always looked after himself, had a heart attack at 40. It shows the importance of picking the right grandparents (that is, heredity) if you want to live a long life.

What happens in the body to cause heart disease? The basic injury or change occurs in the arteries, which carry blood from the heart through the body. Arteriosclerosis is any disease of the arteries, and atherosclerosis is a type of arteriosclerosis which leads to heart disease. The exact origin of the disease is unknown, but many factors are involved and it is a general process that might even begin at birth in some. It begins by some injury or erosion to the arterial wall. This could result from a virus; high blood pressure, which causes the blood to smash against the artery wall at high speeds, particularly where the arteries divide; chemical compounds from cigarette smoke, or high levels of fats or lipids in the blood.

Often this injury can be repaired, but in atherosclerosis a different sequence of events occurs, as is shown in Fig. 1.9. First the injury triggers a response from cells in the blood causing the accumulation of blood cells, called platelets, which are generally involved in wound healing. The platelets send messages to the muscle cells in the artery to begin to multiply and form fibrous material at the site of the injury. At the same time white blood cells, called macrophages, move in and fill up with cholesterol from the blood, forming foam cells. These foam cells concentrate at the injury and release cholesterol, which is deposited on the injury. This combination of events causes a buildup on the scar of the injury called plaque. Soon calcium is added, which hardens and we have a condition where the blood flow is blocked and the artery loses its elastic properties. This is atherosclerosis.

Remember that the blood carries fuel and oxygen to the cells, and if this is interrupted the cells will starve and finally die. The lack of blood

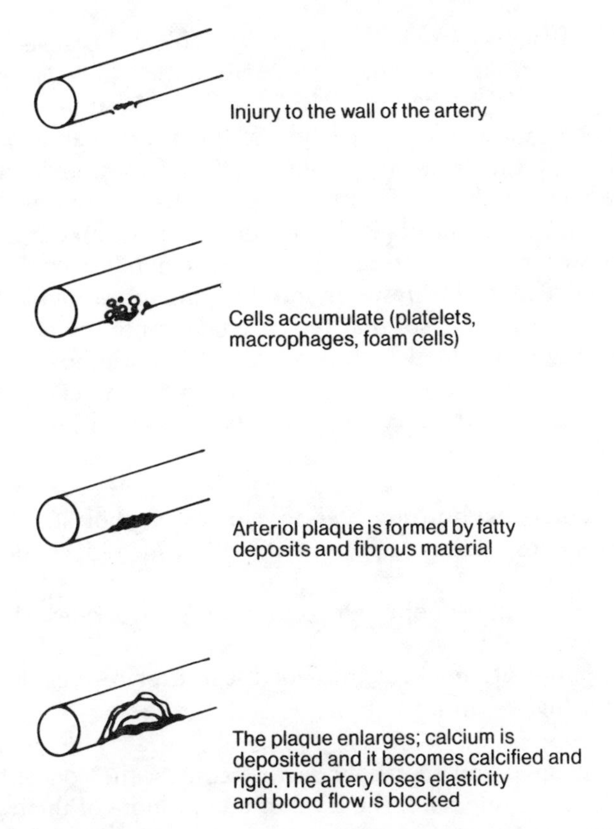

Fig. 1.9. The possible sequence of events in the formation of atherosclerotic plaques.

flow in organs or tissues is known as "ischemia." If the partial blockage occurs in the heart muscle and oxygen supply is reduced, then chest pain called "angina pectoris" develops. As the plaque enlarges, it might cause the blood to clot, thus blocking the artery; this is known as a "thrombosis." If the heart vessels are blocked, we have a "coronary thrombosis," and a "heart attack" or "myocardial infarction" occurs. If the vessels in the brain are blocked, we have a "cerebrovascular accident" or a "stroke," and if the vessels in the legs are blocked, the tissues die and we may have "gangrene." If the "thrombus" or clot breaks away from the artery wall and begins to move it becomes an "embolus." It will continue to move until it gets to a smaller artery and gets stuck thus blocking it. At this new location it is called an "embolism" and can cause the same problems as "thrombosis."

Instead of being blocked, the artery may break because it loses its elasticity. This breakage occurs more easily under great pressure, so if the artery is blocked which increases the pressure or if high blood pressure is present, the weakened portion of the artery may expand like a balloon. This is known as an "aneurysm" and the artery may burst.

What happens in the blood to increase the chances of this occurring? Obviously, one of the factors is the presence of large amounts of lipids, either cholesterol or triglycerides. If you will remember, cholesterol and triglycerides are insoluble in water and thus are not carried in the blood in solution. Remember also, that they are taken into the intestinal wall by bile and then combined with protein to form a lipoprotein called a chylomicron to be taken to the liver. In the liver, various other lipoproteins are formed with varying amounts of both triglycerides and cholesterol:

Chylomicrons—high in triglycerides, lowest in cholesterol
VLDL (very low density lipoprotein)—high in triglycerides, low in cholesterol
LDL (low density lipoprotein)—low in triglycerides, highest in cholesterol
HDL (high density lipoprotein)—lowest in triglycerides, low in cholesterol, highest in protein

Different genetic and metabolic factors control the concentration of each of these lipoproteins. If the level of one or more of them gets very high you are said to suffer from "hyperlipoproteinemia" (hyper = large or great). Six types of different hyperlipoproteinemias have been described (elevations of one or more of the following: chylomicrons, VLDL, LDL, or HDL). These diseases are either inherited, or are a result of some other disease such as diabetes. Estimates vary as to how many people suffer from these disorders, but it is not a high proportion of the population.

Treatment of these diseases includes following a very specific diet, which varies according to the type of disease.

Even if your blood lipid level may not be high enough to categorize you as a specific type, what should you do if your levels are above the normal range? A consensus panel of the National Institute of Health considers triglyceride blood levels below 250 mg/dl as normal. The panel said these levels do not pose a risk if your cholesterol is normal (below 250 mg/dl). However, if your cholesterol level is 250–500 mg/dl, you have borderline hypertriglyceridemia, and you should control your diet, lose weight, stop drinking, decrease sugar consumption, and decrease fat intake. Again, increase fresh fruits, vegetables, grain, cereals, pasta, and bread.

High blood cholesterol has been correlated with the risk of heart disease in many studies. In fact, a recent study by the National Heart, Lung and Blood Institute, of 3,806 men between the ages of 35 and 59 with cholesterol levels above 265 mg/dl showed that treatment with the drug cholestyramine, to lower cholesterol, reduced the risk of heart disease. In fact, they predicted a 2% reduction in risk for every 1% reduction in blood cholesterol. It should be remembered however, that this was a study with a drug not with food.

The American Heart Association recently recommended dietary change for people with high blood lipids. However, they did note that for people with cholesterol levels above the 95th percentile, genetic factors can play an equal or greater role than diet.

The regimen for controlling hypercholesterolemia, hypertriglyceridemia, and generalized hyperlipidemia, was outlined as follows (Food Chemical News, 1984a):

> Phase I: 30% of total calories as fat, 55% as carbohydrate, and 15% as protein. The fat should contain approximately equal amounts of saturated, monounsaturated, and polyunsaturated fatty acids, i.e., each should contribute about 10% of total calories. Complex carbohydrates should constitute the major source of total carbohydrates. Cholesterol intake should be below 300 mg per day;
> Phase II: 25% of calories as fat (with equal amounts of the three types of fatty acids), 60% as carbohydrate, and 15% as protein; 200–250 mg per day of cholesterol;
> Phase III: 20% of calories as fat (with equal amounts of the three types of fatty acids), 65% as carbohydrate, and 15% as protein; 100–150 mg per day of cholesterol.

The report said that "the degree to which intakes of saturated fatty acids and cholesterol are restricted will depend on the severity of hypercholesterolemia, the willingness of the patient to adhere to the diet, and the response of the patient to each dietary phase."

It is noted that, while each person will not need to be willing to proceed to the third phase of the program, "many patients would rather adhere to a strict diet than . . . undergo drug therapy."

Because "hypercholesterolemia frequently has a familial component," AHA recommended that "other family members should develop the habit of eating foods low in cholesterol and saturated fatty acids."

It is interesting to note that problems seem to arise not solely from total cholesterol but more from the amounts of the cholesterol-containing lipoproteins, HDL and LDL. LDL circulates in the blood taking fat and cholesterol to the cells in the body, and it seems to be the main villain. HDL, on the other hand, carries cholesterol from the circulation to the liver thus removing it from the blood. Therefore we are interested in not only lowering total cholesterol, but, in particular, lowering LDL and increasing HDL.

Other factors beside diet seem to elevate HDL. Exercise is an important factor, but a recent study says that exercise works only in conjunction with diet (Food Chemical News, 1984).

Alcohol in very moderate amounts increases HDL, but the risks associated with alcohol consumption do not make this a viable solution. Increased fiber intake will reduce LDL, and decreasing the consumption of saturated fats, as well as total cholesterol, and increasing polyunsaturated fats will decrease blood cholesterol somewhat.

In other words, *all* of these factors must be considered if we are to have an effective chance of decreasing our risk of heart disease. This does not mean that all foods containing saturated fats should be eliminated from the diet. In fact this would be nutritionally unwise for the majority of the population. (Remember there is a segment of the population categorized as one of the hyperlipoproteinemia types that must eliminate nearly all saturated fats and cholesterol from the diet but we are not addressing that segment here.)

Thus, we come back again to our previous recommendations. Good health requires a lifestyle which includes exercise, no smoking, limiting alcohol, controlling weight, decreasing consumption of meats, and increasing consumption of fresh fruits, vegetables, grains, cereals, pasta, and bread. In addition, we might consider utilizing more polyunsaturated products in our visible fats rather than products containing saturated fat.

VITAMINS

Having finished a discussion of the macronutrients that provide energy to the body, it is time to discuss a group of nutrients that do not provide energy and are not involved in body structure, yet are equally as important as the macronutrients. These are the vitamins.

The vitamins cannot be classified together chemically as can the carbohydrates, fats, and proteins, yet they can generally be classified as to function. They are a group of organic (carbon-containing) compounds that, in very small amounts, are essential in the diet in that they aid or allow specific metabolic reactions within the body necessary for normal growth and maintenance of health to take place.

Since vitamins are required in such minute amounts and since their isolation and identification is quite recent (within 50 years for many of them), there is a tendency to view them as somewhat magical. Couple this with an ever-increasing interest in individual responsibility for health and you have a potential problem. Clydesdale (1984) has pointed out that these trends seem positive, since a healthy image and the notion of being responsible for one's health is rather refreshing. However, crediting or blaming one nutrient for nearly everything can force unwise decisions on the naive and gullible consumer. Herbert and Barrett (1982) report that the retail sales of food supplements in the United States of only three companies, Shaklee Corporation, Amway Corporation, and Neo-Life Company of America, exceeded $700 million in 1980. When asked, "In the last 24 hours, which of the following did you do?" readers of *Psychology Today* gave their top two categories as taking vitamins and drinking beer or wine (Rubenstein, 1982). Contrast this with the number of people who avoid sugar because they believe it causes aggressive behavior (although research indicates that exactly the opposite should occur) (Kolata, 1982).

The widespread use of supplements has been confirmed by a Gallup (1982) Study, which shows that 37% of the general adult population takes vitamins. Certainly, there is little argument against the rational use of supplements in reasonable amounts. The concern is with the abuse and the use of nutrients at unproven pharmacological levels as opposed to nutritional levels. A catalogue from Star Professional Pharmaceuticals (1984) for chiropractic professional use advertises a vast array of megadose supplements. This in itself is bad enough, but their advertising is also misleading. For instance, they advertise a "Calcium Plus" tablet for either "prenatal calcium or for those who want to supplement their diet with calcium." However, it is only upon further reading that you find that the tablet also contains 150 mg of ferrous gluconate, which could pose some risk for the consumer who might be unaware that they are taking an iron supplement of about twice the RDA for iron, in a calcium pill.

The use of supplements has been carried to the extreme in a book written by Durk Pearson and Sandy Shaw entitled "Life Extension: A Practical Scientific Approach." *Newsweek* reports that this book, which costs $22.50 and is 858 pages long, has sold over a million copies. This is really the epitomy of "high tech health," they are reported to recommend a mixture of water, vitamins, amino acids, prescription drugs (L-Dopa), and other chemicals such as butylated hydroxyanisole (BHA), a food additive used to prevent oxidative rancidity, to retard aging and ward off illnesses from insomnia to cancer.

Also, they eat plenty of red meat and downplay their reliance on exercise (Salholz and Smith, 1984).

National surveys have indicated that certain segments of our population are not receiving the RDA for some selected nutrients. In these cases the use of single strength (RDA) supplements and/or specific fortification may be appropriate when it becomes evident that food cannot, or will not, do the job. For instance, there may be problems getting the RDA for calcium and iron in women who are dieting and generally intake small amounts of dairy products and meats. However, there is certainly not a need for widespread multisupplement use nor could we ever recommend supplement use over the RDA. Remember, we need what we need and not more, and vitamins only cure that which their deficiency causes.

The history of vitamins is exceedingly interesting. Prior to the twentieth century it was thought that illness was always caused by eating something, and not by the absence of something that belonged in the diet. This absence of certain nutrients from the diet leads to what we now know are "deficiency" diseases.

History has shown us, however, that long before the deficiency diseases were recognized, they were occasionally cured by luck. In 1535, the French explorer Jacques Cartier and his men explored the coast of Newfoundland. This party was decimated by what we now know was a lack of vitamin C. The rest of the expedition was saved by the Indians of the region, who recommended that Cartier and his men drink an extract of spruce needles, which fortunately, were rich in vitamin C. The eighteenth-century Scots discovered that their children could be protected against rickets, a bone-deforming disease, by having them ingest cod liver oil. In eighteenth-century Italy, children suffering from pellagra, caused by a lack of niacin, were treated successfully as soon as they entered special hospitals that ministered to this disease. The cure was probably the fact that the children obtained more meat in these hospitals than they did at home.

The definitive study on vitamin C deficiency (scurvy) was done in the British navy. For years, this ailment had plagued the navy, and it was felt that the great tradition of British naval power would be lost, not because of the enemy, but because of the sickness that constantly attacked the British sailors on long voyages. In 1747, James Lind, a physician to the Royal Navy, carried out what is still considered a marvelously well-controlled clinical investigation. He isolated 12 scurvy victims whose cases were very similar and who had all received an identical diet: water-gruel sweetened with sugar in the morning; fresh mutton broth, puddings, or boiled biscuits with sugar for dinner; and barley and raisins, rice and currants, or sago and wine for supper. Lind divided his 12 patients into six pairs and fed them six different diets. Some of these remedies seem terrible, but some of our food faddists would have us eat things as ridiculous as the supposed cures that these men were fed. Four of the pairs received liquid additives such as cider, vinegar, a dilute sulfuric acid mixture, and ordinary sea water. The fifth pair received a remedy "recommended by a hospital surgeon," which consisted of a paste that contained among other things garlic, mustard seed, balsam of Peru, and myrrh. The men in the last team were treated much better: they were each given two oranges and a lemon every day. After six days, one of the men in this last pair was ready to go back to work on board ship; the other was nursing the remaining patients. This experiment showed that citrus fruit contained something that, when absent from the diet, created a deficiency disease. This something, of course, was vitamin C. In order to prevent scurvy from recurring, the British admiralty in 1795 ordered that every British seaman be provided with a daily dose of fresh citrus juice. As a consequence, the Royal Navy's fighting force doubled by the start of the Napoleonic wars. Interestingly, at that time the British called lemons "limes," and as a result the warehouse area along the docks in London's East End became known as "Limehouse" and British sailors acquired the nickname "Limeys."

The man who really led the way in vitamin research, Elmer V. McCollum, almost died in infancy from a vitamin-deficiency disease—again, scurvy. By the time McCollum was a year old, his family had almost given up on his life. But one day, his mother was peeling apples and she fed him some of the peelings to quiet his whimpering. He seemed to like them, and the feeding continued day after day. In a few days, the mother noted an improvement in the child's health and suspected that the apple peelings were helping him. She thought that the improvement was due to raw vegetables, so she began feeding him these, along with the juice of wild strawberries. This diet was an intuitive one, but of course it became a cure for this particular deficiency disease. McCollum eventually obtained his doctorate and began a career in nutrition that culminated in the discovery of vitamins A and D.

Many vitamins function as coenzymes. Remember that an enzyme is a biological catalyst; it helps along the chemical reactions in the body. A vitamin, then, is an "enzyme helper": it aids in the work of the enzyme. A certain number of helpers or vitamins are required for efficient and healthy body metabolism; any more than this amount can clutter up the metabolism. An intake of more vitamins than are required can therefore do more harm than good.

In general, vitamins may be subdivided into two large groups: the fat-soluble vitamins and the water-soluble vitamins. This allows several useful generalizations to be made.

(1) *Sources of the vitamins.* Fat-soluble vitamins, as their name implies, are soluble in fat. This means that they are found more in fat-containing

substances than in water-containing substances. Therefore, a good source of fat-soluble vitamins would be fat containing materials such as oils and meats. On the other hand, water-soluble vitamins are not as likely to be found in fatty substances. Therefore, the best sources of these vitamins are foods that contain more water, such as vegetables and fruits, which, you will remember, contain about 90–95% water and 80–85% water, respectively.

(2) *Stability*. During processing and also during cooking some vitamins can be lost. When cooking vegetables, one should remember that the vitamins in these vegetables are quite often water-soluble. Therefore, it makes sense to cook the vegetables in a minimum amount of water and to utilize the water in which they are cooked as much as possible. If a vegetable is cooked in a large quantity of water and the water is then thrown out, as is often the case, many of the water-soluble vitamins will have been leached out of the vegetable into this water that is thrown out. Not too long ago, it was common to drink this cooking water or use it as soup stock. In fact it was called "pot-likker" in the southern United States.

The Chinese cook most of their vegetables by stir frying. That is, they place the vegetable in a small amount of oil and cook it very rapidly. The water-soluble vitamins are not leached out because no water is used in cooking, a minimum of heat is applied over a very short time so that fewer vitamins are destroyed by heat, and the oil used for cooking the vegetable may add some fat-soluble vitamins.

Fat-soluble vitamins in a food, on the other hand, are not lost when the food is cooked in water.

(3) *Safety*. Since by definition fat-soluble vitamins are soluble in fat they can accumulate and be stored in the body's fatty tissues over a long period of time. This has some safety implications: whereas water-soluble vitamins, when taken in excess, are normally removed immediately from the body through the urine, fat-soluble vitamins may be stored and accumulate to toxic levels. The overingestion of vitamins A and D is particularly dangerous.

It is interesting to note that you will find foods such as meats, eggs, milk, and liver listed as good sources of some vitamins and should be eaten. Yet you may be told not to eat these foods because they are high in saturated fat and cholesterol. Those people categorized as hyperlipidemic should avoid many, and at times, all of these foods. However, for the majority of the population, a varied diet is the best and safest diet, with moderate amounts of these foods as an important part of this diet.

The Water-Soluble Vitamins

There are nine water-soluble vitamins, vitamin C, and eight vitamins in the so-called B complex of vitamins as follows:

Thiamin (B_1)	Pantothenic acid
Riboflavin (B_2)	Folic acid
Niacin	Vitamin B_{12}
Biotin	Vitamin B_6

All of these vtamins are chemicals and as such are equally effective whether they are derived from "natural" sources, such as bacteria or yeast, or by synthesis in the laboratory. Their function depends on their chemical structure, and therefore all forms must be the same in order to function.

Vitamin C. Also known as ascorbic acid, vitamin C plays several important roles in the body, and probably all of its functions are not as yet understood. However, the amount required to produce good health in a population is known; the RDA of vitamin C for adults is 60 mg per day.

Vitamin C is very important for the synthesis of "collagen," which is the major protein in the connective tissue of the body. As the name implies, connective tissue connects tissue in such body parts as bones, cartilage, teeth, skin, and blood vessels. It also plays a role in amino acid metabolism and may be important in compounds involved with the nervous system. Capillary strength and fragility depends on vitamin C, and it helps the body to absorb iron. It may be involved in other reactions in the body, but its importance is obvious from those already mentioned.

As noted previously, a severe deficiency of vitamin C can result in scurvy. A very common disease 150 years ago, scurvy is characterized by swollen gums, painful joints, and spots under the skin. These spots are tiny hemorrhages caused by the breaking of the blood-vessel walls. These symptoms are accompanied by weight loss and muscular weakness, and the disease eventually leads to death unless treated. Today, scurvy has virtually disappeared.

Because vitamin C is involved in many processes in the body, it is often recommended at very high levels. However, such attempts use vitamin C as a drug rather than a vitamin, a role for which it has never been tested. In this regard, the RDA committee has said, "since many of the claims for significant beneficial effects of large intakes of ascorbic acid have not been sufficiently substantiated, and since excessive intakes may have some adverse effects, routine consumption of large amounts of ascorbic acid is not recommended without medical advice."

Ascorbic acid is the most unstable of all the vitamins. It is easily destroyed by oxidation, which is speeded up by heat and alkali. In other words, when vitamin C is exposed to oxygen under conditions of heat, as in cooking, and a non-acid environment, the vitamin is easily destroyed. Fortunately, an acid medium aids the stability of vitamin C. Fruit juices, which are acid products, lose little, if any, of their vitamin C value, even after processing.

The best sources of ascorbic acid are fresh, frozen, or canned citrus fruits. such as oranges and grapefruit, or their juices. But these are by no means the only sources. Strawberries and cantaloupe are rich in vitamin C, as are cabbage, green peppers, turnips, tomatoes, and potatoes. Unless it is fortified, apple juice, a common breakfast drink, is very low in vitamin C. The following all provide about one-half the recommended daily allowance of vitamin C:

1 cup cabbage
one-quarter of a large grapefruit
one-half of a large orange
¾ cup tomato juice
⅓ cup orange juice
1 oz green pepper
1 cup turnip
2 oz fortified cereal

Remember that vitamin C is a water-soluble vitamin and that care should therefore be taken in preparing some of the foods that contain it. For instance, boiling potatoes with their skins on retains more of their vitamin C value, since the water cannot reach the interior of the potato as easily. For the same reason, all vegetables should be cooked in as large pieces as possible, in as little water as possible, and for as short a time as possible. It should also be noted that when foods such as tomatoes, tomato juice, and citrus-fruit juices are canned, they retain most of their ascorbic acid because of their high acidity.

The B Vitamins. The B vitamins are chemically unrelated except for the fact that they contain nitrogen and are not proteins. They cannot be made by the body and therefore must be ingested, with the exception of niacin which can be made from the amino acid tryptophan.

They all act as co-enzymes in the body and five of them, thiamin, riboflavin, niacin, biotin, and pantothenic acid, are involved in energy production. Two others, folacin and vitamin B_{12}, are necessary for red blood cell formation, and the last, vitamin B_6, has other functions.

Thiamin. Like all the B vitamins thiamin functions as a coenzyme or enzyme helper in the body, allowing the body's metabolism to function fast enough and well enough to sustain health and life. Thiamin's main role as a coenzyme is in carbohydrate metabolism, ultimately producing glucose for energy.

A severe deficiency of thiamin produces the disease known as beriberi (polyneuritis). This disease was common in certain areas of the Far East among people who subsist mainly on polished, white rice and little else in the diet. The more affluent people who ate white rice as part of a varied diet did not suffer from a thiamin deficiency. Beriberi affects the nerves, causing a paralysis in the legs giving the afflicted person a peculiar gait. In some cases, the victim's heart enlarges, the heartbeat slows down, and the appetite and consequently the weight decreases. In the past, beriberi was cured by feeding the victim whole rice, including the bran, instead of white rice. These days, white flour and rice are usually enriched. Many people had died from beriberi in the Philippines, but in 1952 enriched rice was introduced commercially in two provinces of that country. The Food and Agricultural Organization (FAO) and the World Health Organization (WHO), evaluated this program and found initially that enriched rice did not completely prevent the deficiency, but further technological advances, along with acceptance of the enriched rice, have virtually eliminated the disease. In North America, beriberi is virtually unknown due to both the availability of a large variety of food and to the enrichment programs that have been introduced by legislation and the food industry.

Thiamin, a water-soluble vitamin, may be lost or destroyed when cooked in water.

Often, a daily vitamin requirement can be satisfied by a single food source that is rich in the vitamin. For instance, a normal serving of orange juice supplies enough vitamin C to prevent scurvy, a teaspoon of cod liver oil supplies enough vitamin D to prevent rickets, and a large serving of green or yellow vegetables satisfies the need for vitamin A. This is not the case with thiamin. Although there is some thiamin in many foods, there is no single natural or unfortified source that can provide all the thiamin needed daily for health. The RDA for adults is 1.4 mg per day for males and 1.0 mg per day for females. Thiamin must be obtained from many foods, fortified foods, or from a diet supplement. The many foods that one normally eats daily will provide one's daily requirement of thiamin, even though the amount of thiamin in the individual foods is small. Enriched cereals, flour, and bread, potatoes, green peas, dried peas, beans, and milk all contribute to the thiamin requirement. Meat is a good source of thiamin, and pork products are the best source. Half a slice of ham or one and a half sausages contains the same amount of thiamin as four slices of enriched or whole-wheat bread, and one cup of cooked soybeans contains as much as six slices of bread. Organ meats such as kidney and heart, are reasonably rich in thiamin, although these are rarely eaten these days.

Riboflavin. Riboflavin acts as a coenzyme in the transport of electrons in the body (oxidation–reduction reactions) and as such is necessary for the metabolism of all the macronutrients, protein, fat, and carbohydrate for energy.

It is a very stable part of the B complex, but it is still subject to the problems associated with water solubility and alkaline solutions. Riboflavin is destroyed by light when in a liquid form; therefore, an excellent source of riboflavin such as milk should not be left in the sunlight for any length of time if its riboflavin content is to be preserved.

The other vitamins mentioned thus far have been used to cure a particular deficiency disease. Although riboflavin does not cure a particular disease, it helps in the cure of pellagra and other diseases. Moreover, it has a beneficial effect on certain sores that appear around the mouth and nose, and it is necessary for growth.

The adult RDA for riboflavin is 1.6 mg per day for males and 1.2 mg per day for females. Variety meats and organ meats are excellent sources of riboflavin. Milk and chesse are also very good sources. One-half cup of milk contains 0.2 mg of riboflavin; the equivalent amount may be found in the following servings of food:

¼ cup evaporated milk
6 oz salmon
2 tablespoons dried milk
2 cups broccoli
¼ oz kidney
one and a half eggs
2 oz cheddar cheese
½ cup cottage cheese (4% fat)
2 cups cooked soy and other beans
½ cup yogurt
2 oz fortified cereal

In general, riboflavin is fairly stable when processed. It is not destroyed by heat, but it may be leached into the cooking water and then lost if the water is thrown away. Little, if any, riboflavin is lost in baking or broiling. Some foods are fortified with riboflavin.

Niacin. The vitamin niacin, is really two substances, nicotinic acid or nicotinamide, and in these forms acts as a coenzyme in the use of carbohydrate, fat, and protein for energy. Interestingly, niacin can be made by the body from one of the essential amino acids, tryptophan, which you will remember is a component of high quality protein such as exists in animal and dairy products.

Niacin is a specific remedy for pellagra, a deficiency disease marked by dermatitis, diarrhea, and dementia. Pellagra was prevalent in many countries of the world in the early part of this century, and has always been associated with poor diets. At first, it was thought to result from a deficiency of protein, since one of the essential amino acids, tryptophan, is the precursor of niacin. Niacin was originally called the pellagra-preventing (p-p) factor.

The adult RDA for niacin is 18 mg per day for males and 13 mg for females. About one-half of these amounts could be obtained from the following:

1½ cups cooked soybeans
4 oz mackerel
3 oz beef or pork
2 oz chicken meat only
4 oz haddock
¼ cup peanuts, shelled
8 slices wholewheat or enriched bread
½ cup sunflower or sesame seeds
2 servings of fortified cereals

Biotin. Biotin is a coenzyme which aids in transferring carbon dioxide in energy-producing reactions, among others. It is widely available in liver, chicken, vegetables, nuts, and is produced by some bacteria, but is not readily available from meat and fish. There is no RDA for biotin but the RDA Committee suggests an intake of 100–200 micrograms per day for adults and estimate that normal intake is about 300 micrograms per day. Deficiency can produce dermatitis, loss of appetite, and depression, although it is rare.

Interestingly a compound in raw eggs called avidin will prevent absorption of biotin from the intestine, so a deficiency could occur from the consumption of raw eggs in both humans and animals.

Pantothenic Acid. Pantothenic acid functions as a component of coenzyme A (CoA). CoA is involved with nearly all energy-producing reactions. There is no RDA for this vitamin, but an intake of 4–7 mg per day is suggested. Average intake is estimated to be between 10 and 20 mg per day. Deficiency is unlikely since all foods contain some pantothenic acid. Good sources are mushrooms, liver, eggs, yeast, heart, peanuts, and cauliflower. Fruits contain very little pantothenic acid.

No dietary deficiency is known and induced deficiencies produce a variety of symptoms, which may or may not be related to dietary deficiencies. Intake above 10–20 grams will cause diarrhea.

Folacin. Folacin is also known as folic acid or pterylmonoglutamic acid and functions as a coenzyme in the transfer of single carbon units in the chemical factories of the body. It is essential for the production of a group of compounds called nucleotides. Without thymine, one of these nucleotides, DNA cannot be made and normal red blood cells cannot be produced because of improper cell division. Thus, a deficiency causes abnormally large red blood cells, a condition known as "megaloblastic anemia." Other deficiency symptoms might include retarded growth and sores on the tongue.

The RDA for adults is 400 micrograms per day. Good sources are liver, citrus fruits, yeast, spinach, lettuce, other green vegetables, whole grains, cottage cheese, and some nuts and nut products like peanut butter.

Vitamin B_{12}. Vitamin B_{12} contains cobalt, a mineral, which makes it quite different from other vitamins. As a result it is also known as cobalamin. It functions as a coenzyme in all cells of the body and like folacin is necessary for nucleotide and DNA production, which allows normal cell division. Therefore, a deficiency causes formation of enlarged red blood cells and megaloblastic anemia, as does folacin. It is also important in proper functioning of the nervous system. Folacin will cure the anemia resulting from vitamin B_{12} deficiency, but it will not cure the nervous system disorders. Therefore, it is important to determine what the deficiency is before treating because the attempt to self-prescribe folacin could mask a lethal B_{12} deficiency.

Vitamin B_{12} is absorbed only at the end of the ileum (Fig. 1.1) and is required only in small amounts; the RDA for adults is 3 micrograms per day. It may be obtained only from animal products including meat, poultry, fish, milk, cheese, yogurt, and other dairy products. It cannot be obtained from plant products unless they have been contaminated with human or animal excrement. Strict vegetarians should be sure to take a supplement.

Vitamin B_6. Vitamin B_6 exists as three forms: pyridoxine, pyridoxal, and pyridoxamine, which are coenzymes important in the manufacture of protein in the body. A deficiency could result in a deficiency of specific proteins such as heme (the protein portion of hemoglobin which is the iron-containing molecule in red blood cells that carries oxygen throughout the body). These proteins are also important in the nervous system and in the immune system. A deficiency could cause a type of anemia characterized by abnormally small red blood cells, as opposed to the anemia caused by a deficiency of folacin or vitamin B_{12}. A deficiency could also result in problems with the nervous and immune systems; the vitamin also plays a role in the release of glucose from glycogen and may be involved with essential fatty acids metabolism. The adult RDA is 2.2 mg per day for males and 2.0 for females, and it may be obtained from a variety of foods including meats, organ meats (liver), beans, potatoes, and other vegetables.

Overview

Since the B vitamins and vitamin C are water soluble there is no great concern over their toxicity. However, whenever the body is overloaded with any chemical, a potential problem exists. For instance, doses of vitamin C in excess of 1 g (1000 mg) per day are thought to cause kidney stones and gout in susceptible individuals. Some scientists believe excess vitamin C will prevent the action of vitamin B_{12} in the body, although there is disagreement on this issue. There is a possibility that high doses of vitamin C may lead to an increased body need for vitamin C in both adults and the newborn child. Unfortunately, there is a tendency to overdo vitamin C consumption, since it has been claimed that excess amounts will cure everything from colds to cancer. Unfortunately, experiments have been unable to support these claims, with the exception of the occasional study that indicates that *twice* the adult RDA (120 mg) *may* (not does) reduce the severity of a cold but not the number of colds.

In large amounts, niacin, but not nicotinamide, can cause the skin to turn red, tingle and sting. It can also cause irregular heartbeat and upset stomach. Massive doses, in fact, are thought to cause diabetes and lead to peptic ulcers. Intakes of vitamin B_6 above 300 mg per day have toxic side effects, and some studies have indicated that doses above 2 g per day have caused neurological disorders.

Vitamins are essential for life in the dosage recommended (RDA) but have potentially lethal effects when used to excess. This is true of any material we ingest, and, to paraphrase Paracelsus, an ancient Greek physician, it is not the compound itself that is the poison, but the amount. Also, it should be evident that a variety of foods will give us the greatest assurance of obtaining all the vitamins we need. For instance, if we omit animal products from the diet we will not get vitamin B_{12}, and if we omit vegetables and eat only meat we will not get folacin. So, once again we see that all foods eaten moderately have an important place in the diet, but the exclusion of any food group from the diet has the potential for risk.

The Fat-Soluble Vitamins

There are four fat-soluble vitamins:

Vitamin A
Vitamin D
Vitamin E
Vitamin K

These vitamins have certain properties in common. They are not soluble in water, and they are not destroyed by heat. They can be stored in large amounts in the liver, where they form a valuable reserve to be used in time of need. They are absorbed from the intestinal tract with the help of bile salts. Any interference with the production of bile salts will cause a decrease in the absorption of fat-soluble vitamins. The use of mineral oil, either in salad dressing or as a laxative, prevents some absorption of fat-soluble vitamins, and this substance should therefore be used only if a physician orders it. The fat-soluble vitamins are associated with fats and oils in food and thus are obtained from these foods.

Vitamin A. Vitamin A exists in three forms in animals, retinol, retinal, and retinoic acid, and as carotenoids in many fruits and vegetables. The carotenoids, which are orange yellow pigments, are said to be provitamins because they are converted into the vitamin in the body. There are actually a number of nutritionally useful carotenes in plants, but the most effective is beta-carotene. Not all the carotene in a plant is converted to vitamin A; in fact as little as 25% may be converted. For this reason, the concentration of vitamin A is expressed in international units (IU) or retinol equivalents (RE), which takes into account the actual amount of vitamin A available. Vitamin A is heat-stable and is thus not lost during cooking and processing, but it may be destroyed by exposure to oxygen or to sources of ultraviolet light, such as sunlight. Fats tend to go rancid under the same conditions, so when a fatty material such as meat smells rancid, chances are good that most of its vitamin A value has been lost.

How often have you heard the expression, "Eat your carrots—they'll make you see in the dark"? Unfortunately, neither carrots nor vitamins will allow anyone to see in the dark. However, a deficiency of vitamin A certainly impairs dark adaptation in humans; that is, it prevents one from adapting from light to dark as quickly as one should. This is one of the first signs of vitamin A deficiency in humans. Other symptoms are dry skin and tiny lumps forming in the corner of the eye known as Bitot's spots. If the deficiency is very great over a long period of time, the eyelids may become swollen and sore, and the eye disease xerophthalmia develops, which may ultimately lead to blindness. In the developing world some 100,000 children per year suffer blindness from vitamin A deficiency.

Vitamin A is also necessary for the production and secretion of mucus in the body. Mucus is a substance that keeps the epithelial tissues of the body soft and flexible (examples of epithelial tissue are the skin, the intestinal lining, the tongue, the tissue on the eyeball). A deficiency of vitamin A stops the production of mucus, and the tissue becomes hard and is said to "keratinize" and is easily infected. Vitamin A is also important in the growth and development of bones and teeth and in normal hormone production and reproduction.

The adult RDA for vitamin A is 5000 IU or 1000 RE for males and 4000 IU or 800 RE for females.

The richer sources of vitamin A are fish liver oils and the livers of all animals. The older the animal, the greater the amount of vitamin A in its liver and, therefore, the richer the source. Beef liver is a better source than calf liver. Milk fat contains both vitamin A and its provitamin carotene, and so does egg yolk. The proportions vary, partly due to the food eaten by the animal and partly due to the breed of the animal.

The richest sources of carotene, or provitamin A, are green and yellow vegetables. However, the amount of carotene available for absorption varies. Experimental evidence has shown that in some vegetables only 24% is available, whereas in others 74% is available. This availability is due to the type of provitamin present.

Other sources are sweet potatoes, broccoli, carrots, eggs, apricots, winter squash, spinach, cheddar cheese, bananas, lima beans, yogurt, and creamed cottage cheese.

On a varied diet the RDA is relatively easy to obtain, but in developing countries and in some segments of the U.S. population there is some difficulty.

Although they are often recommended by self-styled healers, supplements in excess of the RDA are dangerous. Some Arctic explorers have died from vitamin A toxicity after eating the liver of a polar bear. As Best and Taylor (1966) have noted

> Vitamin A given in excessive amounts causes toxic reactions. Adults who ingested 300 to 500 grams of polar bear liver became severely ill. Headache, vomiting, diarrhea, and giddiness appear promptly. About a week later, desquamation of the skin

> and some loss of hair occurred. The intake may have been about seven million international units since the vitamin A content of polar bear liver may be as high as 18,000 international units per gram. Numerous instances of poisoning have been described in infants and children given excessive dosages of vitamin A in the form of fish liver oil concentrates. Scaly dermititis, patchy loss of hair, fissured lips, skeletal pain, irritability and anorexia were common to all these patients. The insidious onset of the symptoms and prompt response to cessation of overdosage are characteristic. Permanent sequelae are unusual. More cases of vitamin A overdosage than of deficiency have been reported in medical journals in recent years. The condition has been studied in a human volunteer. Nothing is known of the biological properties of vitamin A that will account for these reactions.

You can see that a recommendation of five times the amount of vitamin A required by humans is at best shortsighted. Excess vitamin A cannot be excreted by the body because vitamin A is stored in fat. Therefore it can accumulate until it reaches a toxic level.

Vitamin A has been prescribed for acne, but it has not been found to be effective. Nor, incidentally, has any food including chocolate, milk, sweets or fats been found to cause acne. However, one form of vitamin A, retinoic acid has been used effectively at times as a topical cream for acne. It must be used over a long period of time and causes more acne initially, and it also sensitizes the skin so one must stay out of the sun. This is not a nutrient use but a drug use, and therefore should only be considered under a physician's direction.

Vitamin D. A deficiency of vitamin D results in a disease known as rickets, which has been recognized for centuries but has been cured only as recently as the early part of this century. We know today that vitamin D cures and prevents this disease, which is easily recognized by bow legs, pigeon chest, and enlarged forehead bones. Two forms of vitamin D function in the body, ergocalciferol (vitamin D_2) and cholecalciferol (vitamin D_3). The first comes from a provitamin, ergosterol, in plants and the second from an animal source, 7-dehydrocholesterol, which is made from cholesterol. Both these substances are converted to their active forms on exposure to sunlight.

For years it was known that children in the tropics hardly ever developed a vitamin D deficiency, and initially this was not understood. Later it was found that the provitamin 7-dehydrocholesterol is found in the skin of humans, and that when sunlight reaches the skin it changes this 7-dehydrocholesterol to active vitamin D. The blood then carries the vitamin D to the liver, where it is stored for future use. Fog, rain, and cloudy weather, of course prevent sunlight from reaching the skin. Also, people who do not live in a tropical climate must wear sufficient clothing to protect themselves from the rigors of the weather, and, thus, they are not exposed to the sunlight necessary to convert the provitamin into vitamin D. Therefore, vitamin D must come from foods or food supplements.

Rickets seems to be caused by the inability of the body to absorb calcium and phosphorus in the right proportions from the intestinal tract. In other words, there can be plenty of calcium and phosphorus in the food ingested, but without vitamin D much of these two minerals are lost. Therefore, to prevent rickets, food that contains calcium and phosphorus should also contain vitamin D. One can now see the great rationality in fortifying milk with vitamin D. Milk contains an abundance of calcium and phosphorus naturally, but it does not contain vitamin D. Since all three nutrients are required for the prevention of rickets, it is logical to fortify milk with vitamin D so that the body may obtain all three nutrients at the same time. We have already noted that the use of enriched rice in the Philippines helped prevent beriberi, and now we see that rickets is virtually unknown because of the fortification of milk with vitamin D and the use of vitamin D supplements.

Very few foods contain vitamin D naturally. Certain fish, such as herring, mackerel, salmon, and sardines, contain a fair amount, and egg-yolk contains some, but other foods contain either none or a negligible amount. Because of this, it has been necessary in the past to consume fish liver oil concentrates as a supplement in order to obtain the required supply of vitamin D. The fortification of milk with vitamin D has, of course, eliminated to a great extent the need for fish liver pills.

Excessive vitamin D may also be toxic (Best and Taylor, 1966). The first signs of toxicity are digestive disorders (vomiting and diarrhea) with loss of appetite and considerable loss of weight. Kidney damage finally results in death. Excessive doses of vitamin D cause hypercalcemia and calcification of the joints and soft tissues, especially the kidneys, large and medium sized arteries, heart, lungs, bronchi, pancreas and parathyroid glands.

Vitamin E. This vitamin has been described as a cure looking for a disease. Four forms of vitamin E are known to occur naturally. Of these, alpha-tocopherol is present in natural polyunsaturated fatty acids as an antioxidant. Vitamin E is used by the food industry to prevent oxidation and consequent spoilage of some food products. In this case, a food additive is a vitamin. Chemically, vitamin E exists as alpha-, beta-, gamma-, and delta-tocopherols.

Vitamin E is necessary for reproduction in rats. It was found that on a vitamin E-deficient diet, female rats were unable to bear young, and males became sterile. This research led to the belief that vitamin E was a high-potency vitamin that enhanced sexual capability, and people therefore began to consume larger and larger amounts of it. In recent years this has become a fad again.

Vitamin E may be required in small amounts for the synthesis of heme, the iron containing portion of hemoglobin. However, its major role seems to be as an antioxidant protecting and stabilizing cells, cell membranes, and enzymes.

The adult RDA is 10 mg of alpha-tocopherol equivalents (TE) for males and 8 for females. A human deficiency has never been found, so apparently enough is obtained from the foods we eat. Vitamin E is found in wholewheat bread, wheat germ, and leafy green vegetables. The richest sources are plant oils. Margarine is a very good source, and milk contains vitamin E in direct proportion to its vitamin A content.

Toxicity of vitamin E in humans has not been seen, but with increasing numbers of people self-prescribing large amounts, it could become a problem. Vitamin E can interfere with vitamin K and thus reduce the ability of the body to clot blood.

Vitamin K. Vitamin K has only one function in humans and that is the clotting of blood. This was first noticed in chicks because it cured a hemorrhagic disease characterized by delayed blood clotting. The delay was caused by a lack of sufficient prothrombin in the blood. When the chicks were given vitamin K, however, the amount of prothrombin returned to normal and the blood clotted in the normal amount of time. Prothrombin is formed in the liver where its manufacture is activated by vitamin K.

Plants are the best source of vitamin K. Leafy green vegetables such as spinach and chard contain good amounts, whereas cereals and fruits contain very little. Animal foods are almost void of vitamin K. Fortunately, however, it is synthesized in the digestive tract of many animals, including humans, so we are not dependent on food as the sole source of this vitamin. Bile is required for the absorption of vitamin K, and thus, if the bile ducts are blocked, absorption of vitamin K will be low. Under normal circumstances, sufficient vitamin K is eaten or synthesized in the intestinal tract. However, in surgery, where the clotting time of the blood is most important, extra vitamin K may be given to the patient to prevent hemorrhaging, and vitamin K may be given to the mother before delivery or to the infant at birth. This clotting action is unrelated to hemophilia, which is a genetic disorder.

There is no RDA for vitamin K but a safe and adequate level has been suggested as 70–140 micrograms per day. We obtain much of this from our intestinal bacteria and the rest from food. Deficiency in adults is rare and supplementation is unnecessary. There is also the potential for toxicity as large doses can cause a type of anemia and cause jaundice in infants.

MINERALS

Minerals, like vitamins, are required by the body in small amounts for the maintenance of health and life. However, unlike the vitamins and all the other nutrients we have discussed thus far, they are not organic compounds. We have often wondered how people who advertise "100% organic food" can explain how these foods contain the essential inorganic minerals. Organic compounds if you will remember, are those that contain carbon. The other nutrients have been manufactured by some living form, such as an animal or plant. Minerals are inert material and have been on the earth since its beginnings.

There are 21 minerals recognized as being essential for the human and several more which occur in the body, but their function and essentiality are not known.

The minerals are generally classified into two groups as shown in Table 1.4: (1) the macronutrients (macro = large) minerals which exist in the body at levels greater than 0.005% of body weight and are essential at levels of 100 mg or more per day and (2) the micronutrient (micro = small) minerals which are in the body at levels less than 0.005% of the body weight and are essential at levels of only a few milligrams or less per day.

Minerals, like vitamins, are involved in enzyme reactions as cofactors, but also have several other functions:

Table 1.4. Minerals in the Body

Type	Mineral
Macronutrient more than 0.005% body weight 100 mg or more needed per day	Calcium Phosphorous Sulfur Sodium Potassium Chlorine Magnesium
Micronutrient less than 0.005% of body weight a few mg or less needed per day	Iron Copper Zinc Fluorine Iodine Chromium Cobalt Selenium Manganese Molybdenum Tin Silicon Nickel Vanadium

1. act as cofactors for enzymes
2. constitute a part of essential compounds like vitamin B_{12} and amino acids
3. regulate and balance fluids in the body
4. serve as buffers to regulate acid levels in the body
5. effect transfer of nerve impulses
6. control transfers across membranes
7. constitute major structural components of the body such as in bone
8. necessary for blood formation
9. essential for growth and reproduction

The minerals are a complex group of compounds. They are very reactive, chemically, although they are often thought of as being inert. This is because they can, and do, exist with either positive or negative charges, which effects their ability to be absorbed into the body. The minerals can combine with other compounds in food. Some of these compounds aid in their absorption into the blood and are said to increase the mineral's "availability" or "bioavailability." Other compounds inhibit mineral absorption and thus decrease their bioavailability. Some foods, such as flour and cereals, are fortified with some of the minerals.

Interestingly, the toxicity of minerals is quite high in comparison to other nutrients. For instance, the difference between what is required for good health and the amount that produces toxic symptoms, may only be a factor of 5 or 10, depending on the mineral. This is another good reason to limit intake to the RDA and avoid self dosages. In addition, the absorption of certain minerals is sometimes competitive. For instance, an excess of iron might prevent the absorption of zinc, and vice-versa. In other words, if you take too much of one you might inhibit the absorption of the other and create a deficiency. The body seems to protect itself against overdoses of some minerals because the more there is in the body the less is absorbed from the diet. Conversely, when the level is low, the body absorbs more of the mineral from the diet.

Macronutrient Minerals

Calcium and Phosphorous. Calcium and phosphorus are closely related in the body, being found mostly in the bones and the teeth in the form of hydroxyapatite ($Ca_{10}(PO_4)_6(OH)_2$). Hydroxyapatite is arranged in a crystal structure around softer protein material in the bones and provides strength and rigidity.

Calcium is also involved in blood clotting, transmission of nerve impulses, contraction of muscle, and enzyme action, whereas phosphorous is important in the production of energy as a component of many vitamins and ATP, in absorption of nutrients, as a buffer to maintain acid–base balance in the body, and as a component of DNA and RNA.

Phosphorous is readily absorbed and is available from many foods, the highest amounts coming from the milk and meat groups and high protein foods in general. Carbonated beverages are also often high in phosphorous and deficiency is rare. The RDA is 800 mg per day.

The calcium story is not as simple. The body loses over 300 mg per day in the urine, feces, and perspiration. However, we absorb only about 40% of the calcium we eat, with the body absorbing a greater percentage when the level of calcium is low and less when it is high. This means that we require at least 800 mg per day, which is the adult RDA and more during pregnancy and lactation.

The problem is that calcium is not as widespread in food as phosphorous is. Dairy products are the primary source. Unfortunately, many adults, women in particular, in an effort to lose or maintain weight give up dairy products in favor of foods of fewer calories.

Most of the calcium in the body is in the bones. Bones have two major functions: (1) they provide structure to the body, and (2) they provide a storage reservoir for calcium and phosphorous from which the body can draw when it needs these minerals for other uses. This means that bone, like all other parts of the body, is in a constant state of change. Bone is dissolved or depleted and then built up again, with the total skeleton being replaced about every five years. Because calcium is so important to the body, its level in the blood is regulated by several processes as shown in Fig. 1.10. Parathyroid hormone (PTH) maintains calcium in the blood by stimulating its removal from bone, by increasing reabsorption of calcium in the kidneys to the blood, and by activating vitamin D so that more calcium is absorbed from food in the intestine into the blood. Vitamin D, also known as calcitriol, increases absorption of calcium from food in the intestine into the blood and calcitonin, a hormone secreted by the thyroid gland, lowers calcium in the blood by inhibiting its removal from the bones.

Although dietary calcium is not solely responsible for the maintenance of calcium in bone it does play an important role. Unfortunately, after about 12 years of age women do not consume the RDA for calcium, and the problem becomes worse as age increases. Men consume twice as much calcium as women simply because they eat more food in general and more calcium containing food in particular.

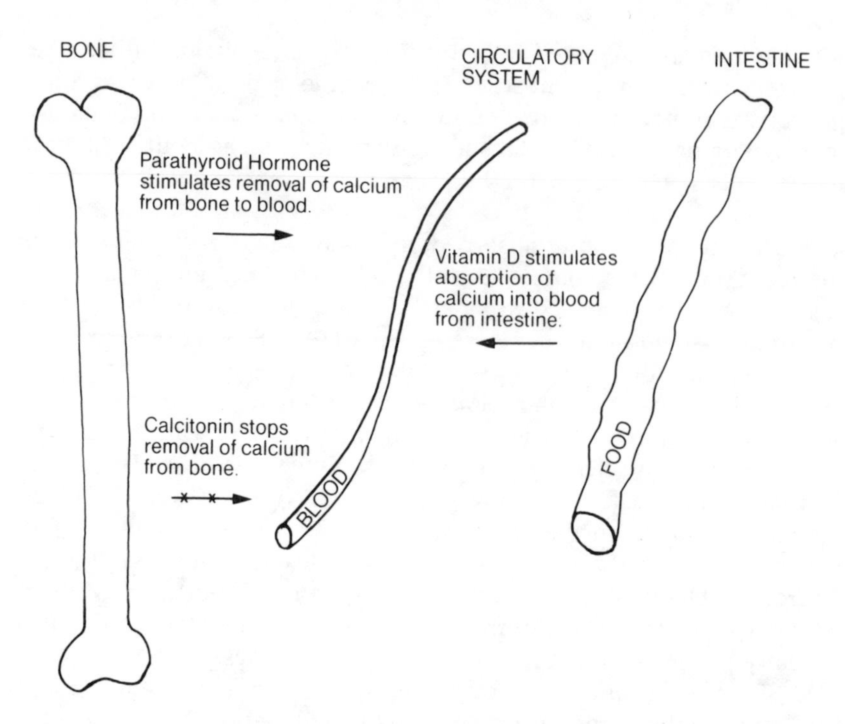

Fig. 1.10. The regulation of calcium in the body between bones, blood and food.

One out of four women (25% of all women) 65 years of age and over will suffer spontaneous fractures of the bones. This is due to a condition known as osteoporosis, in which the bone becomes smaller but is normal in composition, in contrast to osteomalacia where the composition of bone is not normal due to lack of vitamin D or a disease state.

Observations indicate that women in their twenties and thirties are in negative calcium balance, that is they lose more calcium than they take in. This condition is worsened after menopause because of hormonal changes, and some physicians treat osteoporosis with the hormone estrogen. Unfortunately, some studies indicate that this hormone therapy increases the risk of endometrial cancer.

The osteoporosis problem is similar to heart disease in that a number of factors are involved. Hormone changes influences bone density, high protein diets increase calcium excretion in the urine, and materials in plants such as some fibers, oxalic acid, and phytic acid prevent the absorption of calcium. Lactose (milk sugar), and some amino acids increase absorption, exercise helps to keep calcium in the bones, genetics are important, vitamin D is essential for absorption, and there are a host of other factors. From these factors it is obvious why dairy products are such a good choice—they contain not only large amounts of calcium but lactose, amino acids, and vitamin D as well, all of which aid in calcium absorption into the body.

Although calcium alone will not cure osteoporosis the fact remains that increased intake of calcium presents little risk and is probably very helpful in preventing the problem. In fact, many scientists advocate 1000–1500 mg of calcium per day for women rather than the RDA of 800 mg per day.

Even the lesser amounts, 800–1000 mg of calcium, are virtually impossible to obtain without eating dairy products (milk, cheese, yogurt, ice cream, etc.) and/or taking a calcium supplement. Those who have a problem with lactose (milk sugar) can eat such products as yogurt, buttermilk, cottage cheese, and certain other cheeses. Most of those with lactose intolerance can use all dairy products in small amounts.

Recently, another interesting issue about calcium has been raised. D. A. McCarron and his group at Oregon Health Sciences University have found that low levels of calcium intake correlate with high blood pressure when large populations are studied. This does not establish cause and effect, but it does point out that the high blood pressure issue may not be a simple one involving sodium only.

Sodium, Potassium, and Chlorine. The elements sodium, potassium, and chloride are involved in at least six important physiological functions of the body:

1. Maintenance of normal water balance and distribution
2. Maintenance of normal osmotic pressure
3. Maintenance of normal acid–base balance
4. Maintenance of normal muscle function
5. Enzyme activation
6. Transmission of nerve impulses

Water mainly exists in the body either in the cells (intracellular fluid) or outside the cells (extracellular fluid). It is essential for good health that the amount of water in each of these areas remain balanced and this is accomplished by sodium, potassium and chlorine.

Sodium and chlorine exist outside the cells and are said to be extracellular. Potassium exists inside the cells and is said to be intracellular. There is a physical law of nature that requires that the concentrations of solutes such as sodium and potassium be equal on both sides of a membrane (Fig. 1.11). This is known as osmosis, and is accomplished by a

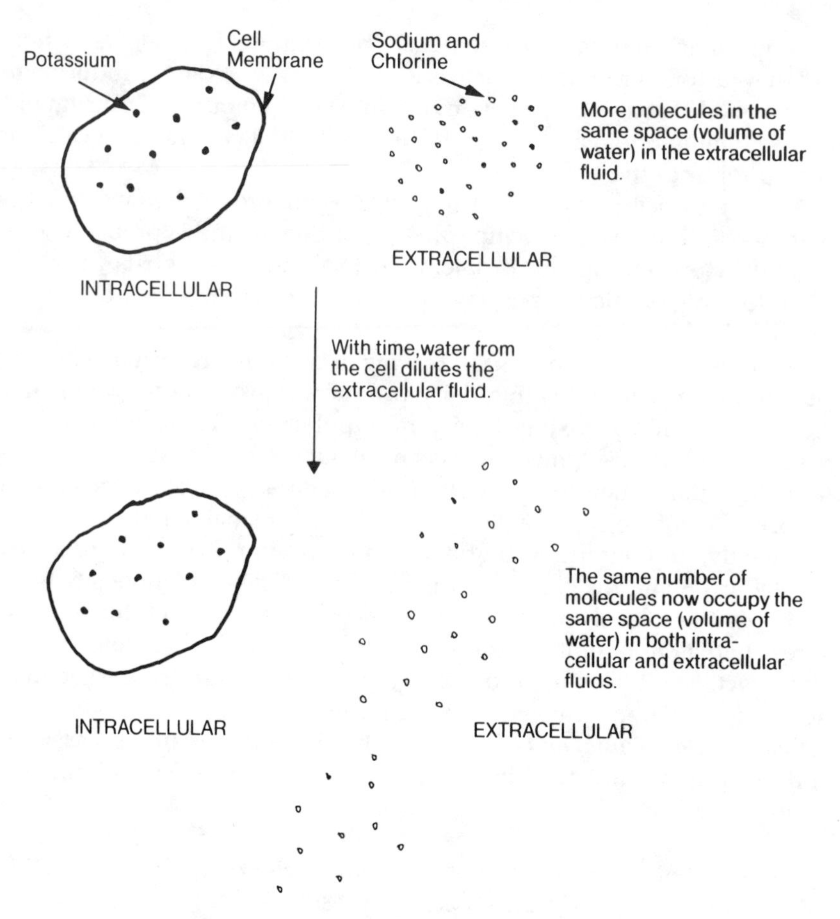

Fig. 1.11. The process of osmosis illustrated by the dilution of extracellular sodium by water from the cell to equalize intracellular concentrations of potassium with extracellular sodium.

force known as "osmotic pressure." This drives water out of the cell and through the cell membrane until the concentration of atoms outside the cell is the same as that inside the cell. Thus, fluid levels and solute concentration are maintained. This may also be accomplished by other solutes such as protein. When we become dehydrated the extracellular water gets low increasing the solute concentration and causes water to be forced out of the cells. A hormone (aldosterone) also causes the kidney to reabsorb more sodium. Another hormone (antidiuretic hormone) causes more water to be retained and signals the brain to tell us to be thirsty—a very neat system.

Since sodium and chlorine control blood volume, it is not unrealistic to suppose that salt (sodium chloride) has an affect on blood pressure. If you increase the amount of water in the blood it will increase the pressure in the circulatory system since it is a closed system.

This points to what would seem to be an obvious cure for high blood pressure, lower salt intake. However, 20% of the population suffers from high blood pressure. Of this 20%, 90% suffer from "essential" or "primary" hypertension from unknown causes. The cessation of salt has not completely cured those suffering from hypertension, and salt therefore is not the only causative factor. The reduction of salt in those who have high blood pressure in their family is a good idea, in any case. The reduction of salt for the general population is a controversial point. Remember, McCarron believes that calcium is also involved in high blood pressure and genetics are extremely important. Weight loss is perhaps the most important first step, and following physician's directions is the next best step.

Acidity is caused by the presence of hydrogen ions that have a positive charge (H^+) and basicity is caused by the presence of hydroxyl ions with a negative charge (OH^-). In pure water, acids equal bases ($H^+ = OH^-$) and we have neutrality. If H^+ exceeds OH^- the solution becomes acid and the reverse is true for basic solutions. The degree of acidity or basicity is measured by a unit called pH. A neutral solution is said to have a pH = 7, an acid solution a pH of from 0 to 7, and a basic solution from 7 to 14.

The presence of certain elements such as chlorine, phosphorous, and sulfur favors the formation of H^+ (acid solutions), while sodium potassium, calcium, and magnesium favor the formation of OH^- (basic solution). Therefore, these elements along with other buffers in the body maintain a neutral pH in the body.

Elements such as sodium (Na) and potassium (K) have a positive electric charge when they are in their ionic form (Na^+ and K^+). Thus, changes in their location into or out of the cell will provide an electric current. This is what happens to cause a muscle to move or a nerve impulse to be transmitted.

The Food and Nutrition Board has suggested an estimated safe and adequate intake for sodium as 1100–3000 mg per day, which is contained in 2.8–4.8 g of salt. This intake is easily achieved by most Americans, in fact, the average intake is 3900–4700 mg of sodium or 10–12 g of salt (3 g from the shaker, 3 g from natural sources, and 4–6 g from processed foods). Interestingly, hypertension is affected more by salt (sodium chloride) than other forms of sodium. Therefore, elimination of the use of the salt shaker would be a help, along with a reduction of the salt in processed foods if it is not used as a preservative.

The estimated safe and adequate daily dietary intake of potassium is 1875–5625 mg per day. It has been suggested that increased potassium along with decreased sodium might reduce the risk of hypertension. Also, remember that on extremely limited high protein diets it is thought that a deficiency of potassium can be implicated in heart failure. Good sources of potassium are meat, fluid milk, many fruits such as bananas and oranges, some vegetables like spinach, and peanuts.

An estimated safe and adequate daily intake of chloride is 1700–5100 mg per day. Chloride is involved with fluid and pH balance and is an important component of gastric juice in the stomach. Nearly all our chloride comes from salt either added or naturally occurring in foods.

It is interesting to note the importance of salt from an historical perspective. Trade routes were established for salt, Roman soldiers were paid part of their wages in salt and thus the word salary (sal = salt). Social status was evident at a dinner table if you were seated above or below the salt, and you could have someone murdered in return for a few pounds of salt. The reason for its attractiveness is not only taste but its preservative qualities for foods; it could also be used as an antiseptic in treating wounds. Today of course, we have the technology to mine as much salt as we want, and it is one of the least expensive spices or flavorants that we can buy.

Magnesium. Back in the early 1600s, people flocked to a place called Epsom in England to drink its bitter water for purgative effects. Today people still drink "Epsom Salts." The major component of Epsom salt is magnesium, an element that activates many enzymes, and, in combination with calcium, phosphate, and carbonate, influences bone metabolism. It is also involved in macronutrient metabolism, protein synthesis, transmission of nerve impulses, and a host of other functions.

About 50% of the magnesium in the body is found in bone and the remaining 50% within almost all the body cells (intracellular).

The adult RDA for males is 350 mg and 300 mg for women. The best sources are dairy products, leafy vegetables, breads, and cereals. Since it does not exist in large amounts in any food, a varied diet will provide only about 300 mg per 2500 calories. This means that most women and many men will be below the RDA for magnesium. Fortified foods may be a viable choice.

An extreme deficiency will result in convulsions and tremors similar to those seen in calcium deficiency.

Sulfur. Sulfur is obtained in the diet mainly from proteins as a component of the amino acids cysteine and methionine. Because it is a component of these amino acids, it is found in most cells of the human body. It is also a component of the B vitamins, thiamin, and biotin and in acetyl CoA, a compound essential for metabolism. There is no RDA or estimated safe and adequate daily allowance.

Micronutrient Minerals

Iron. In 4000 BC, Melampus, physician to Jason and the argonauts, is reported to have put iron fillings in their wine to improve vitality and resistance to spears and arrows. According to television advertising today, taking iron is the way to cure tired blood.

In fact, you may be slightly iron deficient and may need to choose more iron-rich foods or take a supplement, but it is hardly likely that this will give you great increase in vitality. Surveys indicate that iron is one of the problem nutrients, particularly among women in the United States. Its deficiency is also of critical importance in the world, when it has been estimated that there are some 500 million cases of nutritional anemias.

The fact that iron is an essential component of red blood cells is well known. It is an essential part of hemoglobin, the molecule that carries oxygen to all cells and tissues in the body where it may combine with the macronutrients and produce energy. Without iron the red blood cells become small, pale, and fewer in number, finally resulting, over a long period of deficiency, in clinical anemia. Remember that anemias may result from a deficiency of any nutrient that is involved in the production of hemoglobn, an iron-containing protein compound. Therefore, a deficiency of protein, folacin, vitamin B_6, and vitamin B_{12} will cause anemias, as will a deficiency of copper, which is necessary for iron absorption and the snythesis of hemoglobin. There are other anemias not nutrient related, such as sickle cell anemia (genetic), hemolytic anemia (genetics, drugs, vitamin E in infants) where the red cells break, aplastic anemia where red cells are not produced (reaction to certain drugs), and, of course, hemorrhage or any loss of blood.

Iron is also a component of myoglobin, which is a compound similar to hemoglobin that functions as an oxygen reserve in muscles. It is also an important cofactor for many enzymes and is necessary for the manufacture of nucleic acids, DNA and RNA, thus making it important in protein synthesis.

Iron is present in the body in extremely small amounts, considering its importance. The total amount in the body is only a few grams, with about 70% occurring in hemoglobin; the rest stored in the liver, spleen, and bones.

The absorption of iron like some other minerals depends on how much iron we already have in our bodies. If we need iron, we absorb more; and if we do not need it, less is absorbed. This is fortunate, since toxic symptoms could occur at iron levels about 10 times the RDA, but unfortunate, since our intake is limited when our bodies have close to what they need.

Iron exists in two forms in food. Heme iron (iron in hemoglobin) exists in meats and is absorbed the most efficiently of any food iron. Nonheme iron occurs in plant foods and is not absorbed as well. Compounding this problem is the fact that certain fibers and other components of plant foods inhibit iron absorption. Luckily vitamin C increases iron absorption. Therefore, vegetarians should be sure to get enough vitamin C, not only to prevent scurvy, but to help prevent iron deficiency. In developing countries, where the main staple of the diet is plant food anemias are widespread. Any advances in technology that will enhance iron absorption from plant foods will be a great help in improving worldwide public health.

Adult males require only about 1 mg per day and females, due to blood loss in menstruation, about 1.8 mg per day. However, since we absorb only about 10% of what we eat, the adult RDA for males is 10 mg per day and for females 18 mg per day.

It is difficult to obtain this much iron, particularly for a woman. Some estimates indicate that it would require almost 3500 calories per day in a varied diet. Most women would gain weight on that diet.

Red meat and liver (heme iron) are particularly good sources of iron, and poultry and fish are good. Grains, soybeans, and other vegetables have a reasonable amount of iron, but it is not absorbed as well due to the fact that it is nonheme iron and some inhibitors are present. However, the intake of citrus juices (vitamin C) will help that. Many foods are now fortified with iron and some with vitamin C as well. like the breakfast cereals. Some breads and flours are also fortified with iron.

When iron pots and pans were used, some contamination of food with iron occurred, which was beneficial. These pans have been replaced with more modern materials so the benefit from contamination has been lost.

Like all compounds, iron can be toxic in excess amounts. Iron overload can lead to conditions known as hemosiderosis and hemachromatosis, which can lead to liver breakdown and death. The Bantus in South Africa brew and drink a great deal of beer made in iron pots and often get hemosiderosis. This indicates that if you overload the system enough, you can break through the body's unwillingness to absorb iron even when it has enough.

Copper. As was previously mentioned, copper is essential for the absorption of iron and for the synthesis of hemoglobin. It is also involved as a cofactor with various enzymes.

The estimated safe and adequate intake is 2.0–3.0 mg per day. Deficiency is rare but it can result in anemia and bone disease. It is distributed widely in meats, vegetables, cereals, breads, and nuts.

Zinc. Zinc is present in most tissues of the body with the highest amounts in the male reproductive organs and in certain membranes of the eye. It is also a component of insulin, which is essential for the absorption of glucose and subsequent carbohydrate metabolism. It is a cofactor for many enzymes, is indirectly involved with protein synthesis, and is needed for wound healing.

The adult RDA for zinc is 15 mg per day for both men and women. Meats are generally the best and most available source of zinc, and oysters have an exceptionally high level. (This probably accounts for the mythology surrounding oysters and sexual prowess.) In general, high-protein animal foods like dairy products, turkey, and egg yolk are good sources of zinc, along with lima beans, cereals, and bread. Fruits are not a good source.

Plant foods have variable amounts of zinc, since the concentration depends on the amount of zinc in the soil. Plant constituents such as fiber, phytate, and some other materials in bran inhibit zinc absorption, so strict vegetarians should be aware of this. Cases of gonadal dwarfism (immature male reproductive organs) have been reported in areas where very high amounts of unrefined grain products and little or no meat are eaten and zinc is depleted from the soils. This is rare, however. Zinc deficiency could also result in growth retardation, poor wound healing, and a loss of taste perception.

Fluorine. The major role of fluoride in the body is in bones and teeth. Its role in preventing tooth decay was discussed in some detail previously. Fluoride added to water has been shown to decrease cavities in children by 60% and not cause any ill effects. It also may be important in later life in helping to prevent osteoporosis and spontaneous fracture of bone, which affects one out of four women age 65 or over.

Iodine. Iodine is found in minute amounts in the body—approximately 25 mg in the adult. Of this amount, 15 mg are concentrated in the thyroid gland and the rest is found in the blood and in other tissues of the body. The thyroid makes two hormones, thyroxine and triiodothyronine, with some of the body's iodine and stores iodine that is not needed elsewhere. These hormones are necessary for normal metabolism.

When an adequate amount of iodine is not available to the thyroid gland, it does not function properly. Insufficient iodine may result in a simple colloid goiter, a condition characterized by swelling around the neck directly under the chin. This enlargement is due to an effort by the thyroid gland to make more hormones. This kind of goiter is not accompanied by any toxic symptoms, although it may indicate underfunctioning of the thyroid gland. It is important in that is may develop later into a serious condition.

Unfortunately, other thyroid-related conditions may be much more severe. Insufficient thyroid hormones during pregnancy may result in cretinism in the baby, which is characterized by physical and mental dwarfing. If thyroid hormone is given to cretins early enough, they may recover completely. If not, they do not develop normally. Cretinism can be prevented by giving mothers sufficient iodine during pregnancy.

The amount of iodine in food is extremely variable. Sea foods such as oysters, salmon, and seaweeds contain fairly large amounts. Drinking water contains variable amounts. The presence of iodine in a plant depends on the presence in soil. Salt in the United States is fortified with iodine, and as a result deficiencies are less of a problem.

As will be discussed in Chapter 3 some foods contain natural compounds called "goitrogens." These compounds inhibit thyroid hormone production and can cause both goiter and cretinism. Certain parts of Africa where cassava is a major part of their diet have a higher than normal number of such cases due to these compounds.

Interestingly, both a deficiency and an overdose of iodine can cause the same disorders. The adult RDA is 150 micrograms per day for both men and women.

Chromium. Chromium has been recognized as an essential element in recent years, and its major function seems to be to enhance insulin's function in the metabolism of glucose. As a result it is known as the glucose tolerance factor. Chromium in an organically bound form (bound to a carbon-containing molecule not "organically" grown) is thought to be more easily absorbed than free chromium.

Good sources are meat, yeast, cheese, milk, whole wheat breads and cereal, beer, and liver. There is no RDA but the estimated safe and adequate intake is 0.05–0.2 mg per day.

Cobalt. The only known role for cobalt is as a constituent of vitamin B_{12}. Therefore its function and the deficiencies it produces are the same as vitamin B_{12}. There is no RDA nor any estimated intake recommended.

Selenium. The major role of selenium in the body seems to be as an antioxidant like vitamin E. It acts as a component of an enzyme that breaks down materials that cause oxidation to occur. In this way it protects cells and cell membranes.

There is no RDA but the safe and adequate estimated intake is 0.05–2.0 mg per day, with meats, seafoods, milk and cereals being good sources. Large doses of selenium are toxic and deficiency is very unlikely.

Manganese. In animals, manganese has a wide variety of functions including growth, enzyme activity, and lipid and carbohydrate metabolism. However, it has not been as clearly defined in humans and human deficiency is unknown. The estimated safe and adequate intake is 2.5–5.0 mg per day, and it may be found in a wide variety of foods including whole grains and leafy vegetables.

Molybdenum. Molybdenum acts as a cofactor for several enzymes. Deficiency in humans is unknown and its estimated safe and adequate intake is 0.15–0.5 mg per day. It is present in meats, whole grain cereals, and legumes.

Tin, Silicon, Nickel, and Vanadium. Human deficiency is unknown for these four minerals. However, deficiencies have been produced in several animal species, and therefore there is a good chance that they may be essential in humans.

The story on minerals is not over. We are learning more every day, and it is becoming the new frontier in nutrition. This lack of knowledge about either deficiency or toxicity is just one more reason to eat a varied diet. A variety of foods gives us the best opportunity of obtaining all the nutrients we need with minimum risk.

BIBLIOGRAPHY

AMERICAN COUNCIL ON SCIENCE AND HEALTH. 1982. The Cambridge Diet. *ACSH News Views* **4**(3), 5.

BEST, C. H., and TAYLOR, N. B. 1966. The Physiological Basis of Medical Practice, 8th edition, pp. 1435–1436. The Williams & Wilkins Co. Baltimore, Maryland.

CLYDESDALE, F. M. 1979. Nutritional realities—where does technology fit? *J. Am. Dietet. Assoc.* **74,** 17.

CLYDESDALE, F. M. 1984. Culture, fitness, and health. *Food Technol.* **38**(11), 108–111.

CLYDESDALE, F. M. 1984. Fad diets and weight reduction. *Nutrition and the M.D.* **x**(1), 1–3.

HEW. 1977. Morbidity and Mortality Weekley Report, CDC No. 78-8017, Department of Health, Education and Welfare. Center for Disease Control, Atlanta, Georgia.

DEUTSCH, R. M. 1967. The Nuts among the Berries. Ballantin Books, New York.

DEUTSCH, R. M. 1976. Realities of Nutrition. Boll Publishing Co. Palo Alto, California.

FISHER, M. C., and LACHANCE, P. A. 1984. Ranking weight control diets. *Proc. 5th Biann. Kellogg–Canada Nutrition Symp.*

FOOD CHEMICAL NEWS. 1984a. Dietary treatment of high blood fats recommended by heart association. **26**(10), 11.

FOOD CHEMICAL NEWS. 1984b. Plasma cholesterol levels in olympic athletes respond to diet changes: USDA. **26**(10), 12.

FOOD CHEMICAL NEWS. 1984. **26**(5), 14.

GALLUP. 1982. Vitamin use in the United States, Survey VI, Vol. I. The Gallup Organization, Princeton, New Jersey.

HARPER, A. E. 1984. Vitamin supplementation—A skeptical view. *In* Perspectives in

Vitamin Supplementation, Vol. IV, No. 1, p. 5. Vitamin Nutrition Information Service. Hoffmann–LaRoche Inc., Nutley, New Jersey.
HERBERT, V., and BARRETT, S. 1982. Vitamins and Health Foods: The Great American Hustle. George F. Stickley Co., Philadelphia, Pennsylvania.
KOLATA, G. 1982. Food affects human behavior. *Science* **218,** 1209.
KRAUSE, M. V., and MAHAN, K. L. 1979. Food, Nutrition and Diet Therapy, 6th ed. W. B. Saunders Co., Philadelphia, Pennsylvania.
LACHANCE, P. A. 1973. Basic human nutrition and the RDA. *Food Prod. Dev.* **6**(8), 56.
LEVEILLE, G. A., and DEAN, A. 1978. Fats, Diets and Your Health. Nutrition Viewpoint Ext. Bull. E-1192. Michigan State University, East Lansing, Michigan.
RECOMMENDED DIETARY ALLOWANCES. 1980. Food and Nutrition Board, National Academy of Sciences, Washington, DC.
RUBENSTEIN, C. 1982. Wellness is all. *Psychol. Today* **16**(10), 28.
SALHOLZ, E., and SMITH, J. 1984. How to live forever. *Newsweek,* March 26, p. 81.
STAR PROFESSIONAL PHARMACEUTICALS. (1984). Star Winter/Spring Catalog, Plainview, New York.
VAN ITALLIE, T. B. 1980. Diets for weight reduction: Mechanisms of action and physiological effects. *In* Obesity: Comparative Methods of Weight Control, G. A. Bray, (Editor), pp. 15–24. Technomic Publ. Co., Westport, Connecticut.
VASELLI, J. R., CLEARY, M. P., and VAN ITALLIE, T. B. 1984. Obesity. *In* Nutrition Review's Present Knowledge in Nutrition, 5th Edition, pp. 35–56. The Nutrition Foundation Inc., Washington, DC.
WOLF, I. D., and PETERKIN, B. B. 1984. Dietary guidelines: The USDA perspective. *Food Technol.* **38**(7), 80.

2

Nutrition for Athletes

The nutritional requirements of most athletes are satisfied, in general, by the same set of guidelines discussed previously, which include meeting the RDAs or the recommended safe levels for all nutrients. The fact that we have included a separate chapter on athletic nutrition should not imply that there are unique dietary patterns for the health of the athlete. This is not the case, and a healthy person with a good diet will become a healthy athlete with a good diet. However, because so many people are involved in athletics in one way or another, it was decided to include some specific information. From 1970 to the present, participation in amateur softball has increased from 16 to 30 million, golf (15 rounds or more) from 9,700,000 to 13,650,000, and tennis from 10,655,00 to 25, 450,00 (Grandjean, 1982) to say nothing of all the other sports.

Because of the problems peculiar to the performance of most athletic activities, some precautions must be taken. If energy (calorie) levels are met, there should be few problems with the exception of water, iron, and rarely, salt. However, energy requirements often pose a problem in both directions; sometimes the athlete is trying to "bulk-up" or "make-weight" or sometimes "slim-down," which demands increases or decreases in calorie intake.

Also, energy requirements alone do not control energy demands. Like the average person, an individual athlete's energy requirements depend on age, gender, weight, height, body composition, sport, conditioning, clothing worn, playing surface, environment in which the athletic activity takes place and the frequency, intensity, and duration of the event or training session (Grandjean, 1982). Increasingly, both athletes and coaches recognize the potential for a competitive edge, which may be achieved by reducing body fat to a minimal level compatible with good health. This provides maximum strength, endurance, and quickness for each pound of body weight. An average 15- or 16-year-old white male in the United States may be expected to have 14 to 16% body fat, whereas champion high school wrestlers, male gymnasts, and distance runners might have only 5–7% body fat. High school females have 20–22% body fat, whereas participants in highly competitive programs in gymnastics, distance running, cross-country skiing, ballet, or figure skating often reduce body fat to less than 10%.

In weight-matched competitions such as wrestling or light-weight rowing, the athlete with the largest muscle mass for a given weight can have a competitive advantage in strength and endurance. In endurance sports, such as distance running, cross-country skiing, and bicycle racing, fatness greater than a healthy minimum is a disadvantage as it limits speed and contributes to fatigue. Participants in active team sports requiring considerable body movement, like basketball and soccer, are very concerned with the handicap of fatness (Smith, 1976).

Smith (1976) has pointed out that the combination of the advantages of a healthy minimum level of fat in athletic performance, along with the strong negative athletic and moral connotations of over-fatness in our society, create strong pressures on many young athletes to abhor and reduce body fatness. Such self-generated aversion to fatness is commonly reinforced by persons occupying dominant roles in the life of the athlete: coaches, teammates, and parents. As a result, weight loss may continue beyond a healthy minimum and a condition may result which simulates anorexia nervosa in athletes. Fortunately, this condition is more easily treated than anorexia nervosa, and the recovery rate is very high once the condition is discovered and treatment begins under an informed, reassuring physician. It is very important to define an optimum body fat level for the athlete by appropriate measurements, and then have the athlete aim at this healthy level by a combination of training and diet. Some degree of body fat is essential for good health and performance, and to fall below this level is to assume great risk. Obviously, the percentages of body fat quoted previously for champion athletes represent a minimum level. It can be safely assumed that lower levels would pose a danger to health and impaired performance. Also, not all athletes could achieve the levels quoted.

Although there is a great desire to lose weight for many athletes, particularly in weight-matched competition, the greatest problem for young athletes is probably that of obtaining enough calories. Often the athlete is a busy person and time for studies, practice, events, and student activities may replace or at least constrain the time needed for eating. This can lead to dangerously low weight levels and, potentially, ill health.

Total calorie needs vary, as stated previously, but it is not uncommon for an athlete to require 5000–6000 calories per day, which is difficult to achieve without some effort.

WATER

The essentiality and functions of water in the body were discussed in Chapter 1. However, the demands of the athlete may be greater than the average person for two reasons (Smith, 1980).

1. Losses due to daily training and participation in events may cause a loss of water of up to 5% of body weight. This means that an athlete might lose up to 5 pounds of water for every 100 pounds of body weight during rigorous training or competition, which must be replaced.

2. Lean tissue (protein) contains more water than fat tissue does, and athletes generally have more lean as compared to fat than average. As a result, they might have 70% of their weight as water as compared to 60% water for the average person. It is imperative that water losses be replenished as they occur, and to do this one cannot depend on feelings of thirst. Often the intake that will satisfy thirst will not replace the water that has been lost. In fact, water is sometimes not replaced completely until 24 hours or more after the event. Therefore, small amounts of water are recommended at frequent intervals during an event. Dilute sugar solutions may also be used as long as they do not upset the stomach.

IRON

As was explained in Chapter 1, it is difficult for certain populations such as very young children, adolescents, and women of child-bearing age to obtain enough iron. During adolescence, growth imposes extraordinary demands for iron, due to increases in lean mass and the use of iron for new tissue rather than as ferritin, which is the storage form of iron. As a result, the demand for iron of adolescent boys is equal to the demand of adolescent girls with the greater growth rate of boys offsetting the demands due to menstruation in adolescent girls.

This extra demand can normally be met in athletes since their daily food intake is so high. However, judicious choices must be made in accordance with the foods high in iron, or else supplementation should be considered. In either case, if iron deficiency is noted, the use of an intervention program should be considered and checked clinically.

SALT

It is safe to assume that many people in the general population have been told at some time in their lives that they should take salt tablets when involved in heavy physical exertion or when perspiring heavily. In general, there is no need for this, since normal food intake provides more than enough adequate salt, as you will remember from a previous chapter. The average intake of sodium in the United States is 4–6 grams, which translates to 10–12 grams of sodium chloride or table salt. The most agreed upon figure for a minimum requirement is 500 mg (0.5 gram) and the National Research Council Committee on dietary allowances has estimated a "safe and adequate level" to be 1100–3300 mg sodium or 2.8–8.4 g of salt (National Research Council, 1980). Thus, it is clear that the average person is obtaining more than enough salt to meet even heavy exercise in all but the most extreme cases. Indeed, excessive salt intake might create the need for even more water than the excercise itself.

When body water loss and dehydration limits performance in an athletic event, it is not generally due to salt loss in perspiration, which is generally lower in athletes than in the average person, but to the loss of intracellular water required for metabolic energy production. Remember that about 60–70% of adult weight is water. About 40% is in the blood, which circulates through the body, about 35% is inside the cells (intracellular water), and about 25% is between the cells (extracellular water). As exercise commences, water is required inside the cells for the chemical reactions that produce energy. Thus, water moves from outside the cells into the cells. This extracellular water is replaced by water in the blood. In addition to water lost in sweating, this process decreases the amount of water for urine production and thus urinary output is decreased. The overall effect is to decrease both the water in the blood and the extracellular water, thus increasing the concentration of salt in these areas.

In the discussion of minerals in Chapter 1, it was pointed out that one of the functions of salt was to maintain water balance and distribution. That is, as the amount of salt in the blood or outside the cells increases, the body responds by moving water to these areas from inside the cell in order to maintain a constant concentration of salt in all compartments. Since decreased performance is due to a lack of intracellular water, it would not make sense to recommend taking more salt into the body, which would tend to increase further the concentration of blood and extracellular salt, thus pulling even more water from inside the cells. Therefore, the logical recommendation is to drink water, or dilute sugar in water solutions, but not the typical sports drinks, which contain added salts. If it is deemed necessary to add more salt to the diet, it should always be taken with water and not in excess of 1.5 grams salt per liter of water (Smith, 1980).

PROTEIN

Very often athletes are tempted to increase protein intake well beyond the RDA via additional food or pills. Neither method is necessary nor particularly healthy. A healthy diet should have about 50–55% carbohydrate, 30–35% fat, and 12–15% protein. Obviously, if one increases his/her protein intake, one must decrease the percentage of carbohydrates and fat eaten, which upsets the recommended balance. A protein pill is simply a waste of money. The quality of the protein is often suspect, and the amount that can be consumed in a pill is limited. Remember, once the RDA is satisfied, protein is broken down to urea and energy. This increases urinary urea and provides calories the same as carbohydrate and fat do, but you do not get fiber or the other benefits if you get the energy from protein. If the need still exists for extra protein, dry milk powder can be used as an inexpensive supplement.

RECOMMENDED DIETS

At this point it should be reemphasized that the dietary patterns of athletes should be constructed on the basis of the same recommendations made for the average person, as explained in previous chapters. This applies to weight maintenance, weight loss, and weight gain. Fad diets and bizarre eating habits should not be tolerated in the athlete any more than they should be in the average person. Therefore, at the outset, the athlete should aim at eating food from the basic four food groups based on the "four plus one" concept discussed in Chapter 6. Further, there is no place in athletic diets for hormones, diuretics, or cathartics, any one, or all of which can lead to long-term debilitating diseases.

WEIGHT LOSS

In any weight loss diet it is important to attempt to maintain lean body mass while losing a majority of weight in the form of fat. This type of weight loss can be best achieved by a combination of exercise and diet where losses do not exceed 2 pounds per week. Further, a caloric restriction of not less than 2000 calories per day should be followed in order to preserve lean mass. Ideal weight is best achieved during the off-season so that weight reduction exercise programs do not interfere with skill development in season.

WEIGHT GAIN

The young athlete, and the young male athlete in particular, may have a very difficult time in gaining weight. Evidence of the difficulty and the desire to "bulk-up" is clear from the use of anabolic hormones noted in some athletes training for the Olympics. These hormones are not only illegal but are potentially extremely dangerous with regard to sexual potency, atherosclerosis, coronary heart disease, and stroke. In spite of this, they are still used by some ill-informed athletes.

The reason athletes have difficulty in gaining weight is that their energy needs are so high. A young male may require as much as 5000–6000 calories per day simply to maintain weight. Since it takes 3500 calories in excess of need to gain 1 pound, this athlete would have to eat 5500–6500 calories per day to gain a pound a week. This type of caloric intake demands some thought as to content as well as the need to eat more than three meals a day. In order to estimate caloric needs, the athlete who is maintaining weight should keep a record of everything eaten for a week. This will allow a basis on which to devise a diet to provide more calories. Such a diet will necessarily be high in fat, high in low-fiber, low-bulk carbohydrates, and low in high-fiber, high-bulk carbohydrates. Beer does not help much, since even 12 bottles a day (an unhealthy amount) would only provide 1800 calories, and it would fill you up if you could drink that much.

With these constraints it would be wise to use unsaturated fats rather than saturated fats for frying, baking, as spreads, and in salad dressings. Sufficient saturated fats would be obtained from animal products such as milk, meat, cheese, and ice cream, and therefore any additions to food should be unsaturated fats. This would provide a balance in the diet and lessen the risk of increasing blood lipids in susceptible individuals. An important point should be made at this time. Weight gain diets should be used only to attain a goal and only while activity levels are high. Levels of fat in the diet may reach as high as 45% of calories. However, this level is not in keeping with a healthy maintenance diet, which should have only about 30–35% of calories as fat, with continued emphasis on the use of polyunsaturated fats for added fat or oil in the diet. It is instructive to look at the weight of former athletes who have not decreased their intake of food but have decreased their activity levels. Their weight is high, and they look as if their body composition is high in fat. Interestingly, a recent report on U.S. football players who had been listed in *Who's Who in American Sports* indicated that only 65% lived past the age of 50 and the average age of death was 57 (Smith, 1980). As discussed previously, obesity increases the risk of many diseases, and these diseases do not differentiate between former athletes and non-athletes.

A comment on "pills" should be made at this point. There are several types of "energy pills," "protein pills," and other pills on the market that are said to provide some kind of magic energy source for the athlete. These pills will not do the trick. Consider that carbohydrates and proteins provide 4 calories per gram and fat 9 calories per gram. Just think about how large a pill you would need to provide a significant amount of enery. A 10 gram pill would be about the size of 20 vitamin pills! If this pill were 100% fat, which provides more calories for humans than any other source, it would have only 90 calories. Not much, in fact this is only a little more than 1% of daily needs if the daily intake is 6500 calories.

SPECIAL DIETS

There are many special diets the athlete is subjected to and to which the athlete subjects himself/herself. There is a mystique about food and performance which at times has very little to do with scientific fact. Years ago, people ate bear or lion meat as opposed to rabbit meat, for example, before going into battle. Obviously the only difference this made was psychological and not physical, since the meat from all animals is made up of the same proteins, fat, carbohydrates, and water. Today, things have not changed that much. If an athlete breaks a record on a given day, the food he or she ate that day will have special properties attributed to it. The problem is that proponents of special diets often do not realize what makes the diet special. Perhaps the occasion where this kind of thinking is most apparent is the pregame meal, and in line with this the following goals are recommended by Smith (1980) in planning the pregame diet:

1. Energy intake should be adequate to ward off any feeling of hunger or weakness during the entire period of the competition. Although pre-contest food intakes make only a minor contribution to the immediate energy expenditure, they are essential for the support of an adequate level of blood sugar, and for avoiding the sensations of hunger and weakness.
2. The diet plan should ensure that the stomach and upper bowel are empty at the time of competition.
3. Food and fluid intakes prior to and during prolonged competition should guarantee an optimal state of hydration.

4. The pre-competition diet should offer foods that will minimize upset in the gastrointestinal tract.
5. The diet should include food that the athlete is familiar with and that he is convinced will "make him win."

Another diet that has been and is used by a number of athletes is the High Performance Diet. This diet is loosely based on the scientific fact that energy production in the body is related to the length of time over which the energy is demanded. Energy for short spurts is produced in the body anaerobically (without oxygen) by adenosine triphosphate and phosphocreatine (ATP–PC). After ATP-PC are used up, glycogen in the muscle replaces ATP anaerobically (some studies have claimed that higher levels of glycogen can improve performance). Longer events demand the use of aerobically (oxygen demanding) produced energy, which uses carbohydrates and fats.

The high performance diet is based on the assumption that increasing concentrations from 1.75 grams of glycogen per 100 grams of muscle tissue to as high as 4.75 grams/100 grams will significantly improve performance. This is known as glycogen loading and is achieved through diet. Seven days before the event, a low carbohydrate diet is eaten and workouts are strenuous in an attempt to deplete the muscle of glycogen (glycogen is the storage form of carbohydrate in muscle and liver). Three days before the event, the athlete changes over to light exercise and a high carbohydrate diet in order to replace and build up the glycogen to higher than original levels. This diet is for events that last longer than 30 minutes, and there are some potential disadvantages such as muscle stiffness, upset stomach, hormonal changes, as well as unknown long-term effects. In addition, this is a diet which may be less than palatable to some. As a result, athletes often skip the whole regimen and simply go on a high carbohydrate diet for a few days prior to the event. Perhaps this type of diet should be viewed in the same way as any pregame diet. If it satisfies the goals outlined previously, then its use might be considered.

SELF-PRESCRIPTION AND ITS DANGERS

Self-prescription and self-diagnosis are all too common in our society and particularly so in special groups like athletes. An injury or a feeling of ill health to the athlete is much more important in his or her life than the same injury or ill feeling might be to the average person. For instance, a hang nail on the right thumb of a right-handed quarterback the morning of a game could be devastating.

Since both emotional and physical health is so important on the day of an event, it is understandable why the athlete might be tempted to try any scheme to get the "edge" on the competition. Some of these schemes are harmless, others are not. Obviously, common sense will benefit you greatly. If the scheme provides approximately the RDA for nutrients, then you might consider trying it. If it does not and promises great things from impossible doses, then forget it.

BIBLIOGRAPHY

GRANDJEAN, A. C. 1982. Energy requirements of the athlete. Proceedings of the Stokely–Van Camp Annual Symposium, Food in Contemporary Society. May 26–28, University of Tennessee, Knoxville, Tennessee.

RDA. 1980. National Research Council, Food and Nutrition Board Committee on Dietary Allowances. Recommended Dietary Allowances, Ninth Revised Edition. National Academy of Sciences, Washington, DC.

SMITH, N. J. 1976. Food for Sport. Berkeley Ser. in Nutrition. Bull Publishing Co., Palo Alto, California.

SMITH, N. J. 1980. Excessive weight loss and food aversion in athletes simulating anorexia nervosa. *Pediatrics* **66,** 139.

U.S. DEPARTMENT OF COMMERCE, BUREAU OF THE CENSUS. 1983. Statistical Abstract of the United States, 103rd Edition, pp. 234–236. U.S. Government Printing Office, Washington, DC.

3

Food Additives

Food additives! This is probably one of the most emotional subjects relating to our food supply, one of the issues least understood by the consumer, and one of the most difficult to regulate since absolute safety is impossible to demonstrate.

Simply defined, a food additive is any minor ingredient added to a food to achieve a specific technical effect. In broader terms it may also include compounds unintentionally added due to the production, distribution, and/or processing of a food, but not those that appear in a food accidentally. The term food additive does not include pesticides, color additives, new animal drugs, or any substance used in accordance with a sanction or approval granted prior to the effictive date of the food additives amendment to the existing laws in 1958.

In legal terms a food additive, according to the Federal Food, Drug and Cosmetic Act (U.S. Code, Section 321) is a substance "the intended use of which results or may reasonably be expected to result, directly or indirectly, in its becoming a component or otherwise affecting the characteristics of any food . . . ," with certain exceptions. Excluded from the requirements of this act is any substance that is "generally recognized among experts qualified by scientific training and experience to evaluate its safety, as having been adequately shown through scientific procedures . . . to be safe under the conditions of its intended use. . . ." Such substances are commonly referred to as GRAS substances. Generally Recognized As Safe (GRAS) are compounds that have been used over long periods of time by humans without undue effects and which are deemed harmless by qualified scientists. With respect to GRAS substances in use prior to 1958, the basis of the judgment of safety may be either "scientific procedures" or "experience based on common use in food," but after 1958 the basis must be "scientific procedures."

In 1902 Dr. Harvey W. Wiley established what became known popularly as "Dr. Wiley's poison squad." This squad was set up by the

Federal Food, Drug and Cosmetic Act "to enable the secretary of agriculture to investigate the character of food preservatives, coloring matters and other substances added to foods, to determine their relation to digestion and health and to establish the principles which should guide their use." At that time—back in "the good old days"—there were no checks at all, almost anything could be added to foods, and as long as it did not kill someone, these additives did not have to be tested or screened by any governmental agency. Volunteers for this "poison squad" were recruited from among the employees of the Department of Agriculture. Dr. Wiley said, "I wanted young, robust fellows with maximum resistance to deleterious effects of adulterated foods. . . . If they should show signs of injury after they were fed such substances for a period of time, the deduction would naturally follow that children and older persons more susceptible than they would be greater sufferers from similar causes" (Benarde, 1975). Twelve men volunteered. This squad was responsible for taking many dangerous food additives off the market. During the eight decades that have passed since Wiley's time, analytical procedures have increased tremendously. The testing that is done in the food industry today make our food supply abundantly safe compared to what it was in the "good old days."

The Pure Food and Drug Act of 1906 requires that our food supply be safe. The law states that substances (chemicals) may be added to food only under the following two conditions: that the substances be safe for human consumption and that they serve a useful purpose. A 1938 amendment to this act provided the government with the leverage for enforcing the regulations but it contained an inherent weakness. The Food and Drug Administration had to prove the presence of a deleterious substance before it could act against it. However, the color additive amendment of 1960 closed this loophole by stating that no chemical can legally be used in any food until the manufacturer has proved to the FDA that it is safe. This means that the manufacturer has to do the research, has to pay the bills, and has to prove beyond the shadow of a doubt to the FDA that a food additive is safe prior to its use.

As you can see, we have come a long way: We now have regulations that can be enforced in an effort to keep our food supply safe. We should ask, however, how food additives rank among what are thought to be possible dangers to our food supply. H. R. Roberts, a former Director of the Bureau of Foods of the Food and Drug Administration has pointed out that the Bureau of Foods has consistently ranked food additives in the lowest risk category (Roberts, 1981) as has another former director, Dr. Wodicka and a former Commissioner, Dr. Schmidt.

According to their ranking, based on the risk criteria of severity, incidence, and onset of disease, the following food hazards are listed in their order of importance with one being the greatest risk and five the least risk:

1. Microbiological contamination (includes conventional food poisoning)
2. Nutritional hazards [e.g., undernutrition, interactions of foods]
3. Environmental contaminants
4. Naturally occurring toxicants in foods
5. Food and color additives

This classification is agreed upon by most food technologists and food scientists.

HOW FOOD ADDITIVES ARE APPROVED

Prior to discussing the process of approval, a discussion of the Delaney Clause might prove informative. It deals with the possible carcinogenic (cancer-causing) effects of food additives.

The Delaney Clause was added to the Pure Food and Drug Act as a result of an amendment offered by Congressman Delaney in 1958. It established a food additive regulation prohibiting the addition of any cancer-producing substance to human food. Upon initial reading, this appears to be a very admirable proposal. Indeed it is, except that it does not specify the amounts required for a substance to produce cancer. Many substance when added to the body in large enough quantities may cause cancer. For instance, it has been estimated that one barbecued steak contains the same amount of benzopyrene (a potent carcinogen) as 600 cigarettes. Yet, we eat steak and don't see an increase in stomach cancer. In fact, we have seen a decrease, as we will point out later in this chapter.

Much of the scare concerning cyclamates, saccharin, food colors, and other additives has resulted from experiments in which massive dosages of these compounds were given to animals. At these tremendous levels, cancer has at times occurred. However, the dosages are so much higher than we could ingest that it seems incredible to ban them on this basis. The scare has also developed from experiments in which some additives are injected by syringe into the brain, implanted in the uterus, or injected into chick embryos—hardly the way we consume food additives.

As a result of such testing there has been a flurry of activity by consumers and in the Congress to amend the Delaney Clause so that dosage levels as well as both potential risks and benefits are considered.

Testing for Regulated Food Additives

A substance newly proposed for addition to food must undergo strict testing designed to establish the safety of the intended use. A petition must be presented to the FDA that includes the following information:

1. The identity of the additive, its chemical composition, how it is manufactured, specifications necessary to assure its reproducibleness, and any other means necessary to establish what the composition of the additive is.
2. Information on the intended use of the additive, including copies of the proposed labels.
3. Data establishing that the additive will accomplish the intended effect in the food and that the level sought for approval is no higher than that reasonably necessary to accomplish the intended effect.
4. An analytical method capable of measuring the amount of the additive present at the tolerance levels.
5. Data establishing that the intended use of the additive is safe. This requires experimental evidence, which is derived ordinarily from feeding studies using the proposed additive at various levels in the diets of two or more appropriate species of animals.

Are Food Additives and Gras Substances Safe?

There is no way in which absolute safety can be guaranteed, as is the case in any area of life. Premarketing clearance requirements under the Food Additives Amendment to the Pure Food and Drug Act do assure that the risk of the occurrence of unanticipated adverse effects is at an acceptably small level, at least for food additives. Such assurances may not be available in the case of those GRAS substances that were exempted from the need for laboratory testing by the definition in the law. Recognizing this problem President Nixon in his Consumer Message of October, 1969 directed the FDA to review the safety of each item on the GRAS list. This evaluation has been recently completed.

The establishment of premarketing clearances for a particular substance should be understood to be based on the scientific knowledge available at that point in time. Since our scientific knowledge is dynamic and expanding, we must periodically review our earlier decisions to assure that our assessment of the safety of the substances added to our foods remains up to date. Thus, when changes are made in previous clearances these should be recognized as an assurance that the latest and best scientific knowledge is being applied to enhance the safety of the food supply.

Evaluation of the Gras List

As a result of the Presidential directive to review the GRAS compounds a contract was made with the Federation of American Societies of Experimental Biology. A group of qualified scientists was chosen and designated the Select Committee on GRAS substances (SCOGS). These scientists independently evaluated the GRAS compounds and published a series of reports known as "SCOGS" reports.

Each GRAS compound reviewed was assigned into one of five categories based on the conclusions reached about that compound as follows.

Category 1

Conclusion:

There is no evidence in the available information on (name of substance or substances) that demonstrates or suggests reasonable grounds to suspect a hazard to the public when it is (they are) used at levels that are now current or that might reasonably be expected in the future.

Action Taken:

Continue in GRAS status with no limitations other than good manufacturing practice.

Category 2

Conclusion:

There is no evidence in the available information on (name of substance or substances) that demonstrates or suggests reasonable grounds to suspect a hazard to the public when it is (they are) used at levels that are now current and in the manner now practiced. However, it is not possible to determine, without additional data, whether a significant increase in consumption would constitute a dietary hazard.

Action Taken:

Continue in GRAS status with limitations on amounts that can be added to food.

Category 3

Conclusion:

While no evidence in the available information on (name of substance or substances) that demonstrates a hazard to the public when it is (they are) used at levels that are now current and in the manner now practiced, uncertainties exist requiring that additional studies should be conducted.

Action Taken:

Issue an interim food additive regulation requiring commitment within (specified number) days that necessary testing will be undertaken. GRAS status continues while tests are being completed and evaluated.

Category 4

Conclusion:

The evidence on (substance or substances) is insufficient to determine that the adverse effects reported are not deleterious to the public health when it is (they are) used at levels that are now current and in the manner now practiced.

Action Taken:

Establish safer usage conditions as a food additive or remove ingredient(s) from food. Interested parties may subsequently submit a petition establishing conditions of safe use.

Category 5

Conclusion:

In view of the deficiency of relevant biological (and/or other) studies, the Select Committee has insufficient data upon which to base an evaluation of (substance or substances) when it is (they are) used as a food ingredient.

Action Taken:

Insufficient data upon which to base an evaluation.

As a result of this action 297 compounds have been assigned to Category 1, 69 to Category 2, 21 to Category 3, 5 to Category 4 and 30 to Category 5.

WHY DO WE USE ADDITIVES IN OUR FOOD?

The use of additives is not new in the twentieth century. When humans first learned that fire would cook and preserve their meat and that salt (sodium chloride) would preserve it without cooking, the use of additives began. Food colors were used in ancient Egypt. In China, kerosene was burned to ripen bananas; the reason this method succeeded, although the Chinese did not know it, was that the combustion produced the ripening agent ethylene. Flavoring and seasoning were arts in many ancient civilizations, and as a result, spices and condiments were important items of commerce. Columbus sailed for the Indies in search of food additives—that is, spices. As our knowledge of food preservation and technology has increased, our use of additives has also increased.

More than forty functions now served by food additives can be listed. In this discussion, however, it is simpler to group additives into five broad categories: flavors, colors, preservatives, texture agents, miscellaneous substances.

Flavors

Spices and natural and synthetic flavors are used to complement, improve, and enhance the flavor of the foods we eat.

Colors

These are used mainly to give food an appetizing appearance, on the tested assumption that the way food looks has an effect on its palatability.

Preservatives

Spoilage can be prevented or retarded not only by additives but also by physical and biological processes such as heating, refrigeration, drying, freezing, fermenting, and curing. Some of these processes, however, achieve only partial preservation. Additives, therefore, have a role in prolonging a food's keeping qualities—a major need if we are to feed our modern, urban society effectively.

The seriousness of the problem of spoilage is shown by the World Health Organization's estimate that about 20% of the world's food supply is lost in this way. Indeed, shortages of food in many parts of the world could be alleviated by the wider use of preservatives.

Texture Agents

These are used to stabilize and thicken. Many of the newer convenience foods have become practicable only as a result of the development of new and improved emulsifiers and stabilizers.

Miscellaneous

This group is so numerous that we can indicate only a few of the functions they serve. Among other things, they help retain moisture, and they add nutrients such as vitamins and minerals. Moreover, the acids, alkalies, buffers, and neutralizing agents included in this category are required for the quality production of baked goods, soft drinks, and confectioneries.

Do we want to continue to use additives? Only the consumer can answer this. Consumer acceptance, after all, keeps the manufacturer in business. If consumers will accept bread that stales and molds in a few days, oil and fat products that rapidly grow rancid, wine that discolors, frozen products that separate and become watery, and overall spoilage and waste of food to the extent that we would have difficulty feeding our population—then we can eliminate additives.

There is no real difference under the law between those substances we call "foods" and those we call "food additives." Meat and potatoes, for example, would clearly be considered "additives" instead of "food" when served as stew, except that they are more appropriately considered as GRAS substances. Preservatives such as sodium and calcium propionate are produced naturally in Swiss cheese. Citric acid, a widely used food additive, is present in all citrus fruits. On the other hand, many so-called natural foods contain toxic substances. Safrole, a carcinogen, is found in sassafras roots. The *Whole Earth Almanac* recommends the use of sassafras roots as a beverage and to flavor foods. Sassafras is known as the root beer tree, but in 1960 the FDA banned the use of safrole as a flavor in root beer. Oxalic acid is found in spinach and rhubarb, and certain goiter-inducing substances are found in some vegetables. Solanine, a toxic alkaloid, is present in potatoes. Patulin, a natural carcinogen, is found in flour and orange juice. Other natural carcinogens in foods are thiourea and several related chemicals in cabbage and turnips, and tannin in tea and wine. The list is much longer, but the point has been made: alarm about additives may have diverted us from some more serious concerns about food. For intstance, Larkin (1983) in the magazine *FDA Consumer* points out that

> While many peppermint, rose hip, orange, and others of the more usual herbal teas do offer delicious alternatives to two traditional drinks that contain caffeine—coffee and common tea (*Camellia sinensis*)—we cannot conclude from these facts that all herbal teas are safe, nor that it's safe to consume large amounts of any herbal tea over extended periods. In weighing the safety of this practice, it's very important to note a number of cautions. Caution Number One: Some herbs contain the wrong kind of magic, Found among the herbs are nature's most potent poisons. Caution Number Two: We don't know enough about herbal teas to conclude that they are safe. Caution Number Three: Doctoring yourself with herbs can be very dangerous. Caution Number Four: Ne quid nimis—moderation in all things. Caution Number Five: Not all men are created equal, nor women either [genetic variability will produce different reactions in different people]. Caution Number Six: Remember the old, bold mushroom hunter. Among those who gather wild mush-

rooms there is a cautionary proverb. "There are old mushroom hunters. And there are bold mushroom hunters. But there are no old, bold mushroom hunters." Thus if you gather your own herbs to brew a cup of tea be absolutely 100 percent certain that the herb you pick is the herb you seek.

For those interested in herbs, this article also has a list of unsafe herbs from some 28 plant sources, which should not be used in foods, beverages, or drugs.

Most additives have been tested for safety far more often than most naturally or traditionally processed foodstuffs. It is possible that dozens of familiar foods would be banned if all foods were tested as much as additives are, and if Delaney Clause standards were applied to them. With per-capita consumption of additives in the United States being only 139 pounds or so a year, versus about three quarters of a ton for all other foods, the additive threat might by itself be viewed as relatively minor.

Boffey (1982) in an article in *The New York Times* quotes John Higginson, founding director of the World Health's Organization International Agency for Research on Cancer as saying that most cancers are caused by environmental factors. But, by "environment" we do not simply mean "industrial pollution," but also such lifestyle factors as smoking, alcohol, exposure to the sun, and diets. "The only evidence we have seen of a cancer epidemic in North America and Europe is due primarily to tobacco."

A study published in the June 1981 issue of the *Journal of the National Cancer Institute* is also discussed by Boffey (1982). This study by two world renowned British epidemiologists, Sir Richard Doll and Richard Peto of Oxford, ranks the relative importance of the major factors linked to cancer as follows:

Factor	Percent of Causes
Tobacco	25–40%
Diet (except food additives)	10–70%
Infection	1–10%
Sexual factors	1–13%
Occupation	2–8%
Alcohol	2–4%
Geography (sun, etc.)	2–4%
Environmental pollution	2–4%
Medicine, medical treatment, X-rays	0–1%
Food additives	−5 (may prevent)–2%
Industrial products	1–2%

From this list it is obvious that a single number cannot be assigned to any cause. However, it does provide some measure of the relative contributions of each of the factors. Interestingly, food additives are listed as a preventative in some cases. This is consistent with evidence found that two antioxidants, butylated hydroxytoluene (BHT) and butylated hydroxyanisole (BHA), prevented stomach cancer in mice. In fact, one researcher suggested that the addition of these compounds to breakfast cereals coincides with the decline in stomach cancer in the United States. Other antioxidants such as ascorbic acid and selenium have also been shown to be helpful in some cases. However, this does not mean that the consumer should start taking supplements of these compounds. The amounts in foods are known to be safe, but excessive usage could lead to serious problems. They have not been tested at high levels of phamacological use nor is their effect proven. Solholz and Smith (1984) in an article in *Newsweek* reported that Durk Pearson and Sandy Shaw in their book "Life Extension: A Practical Scientific Approach" recommend taking a mixture of water, vitamins, amino acids, prescription drugs (L-dopa) and other chemicals such as BHA to retard aging and ward off illnesses from insomnia to cancer. Such recommendations are potentially dangerous, and the benefits are unproven, so it make little sense to follow these recommendations.

The general issue of diet and cancer is a confusing one. A committee of the National Academy of Sciences (Committee on Diet, Nutrition and Cancer, 1982) issued a report on Diet, Nutrition and Cancer in which they concluded "The evidence reviewed by the committee suggests that cancers of most major sites are influenced by dietary patterns. However, . . . the data are not sufficient to quantitate the contribution of diet to the overall cancer risk [n]or to determine the percent reduction in risk that might be achieved by dietary modification." Nevertheless they did make some general recommendations about diet and cancer which will be discussed in Chapter 5.

It should also be remembered that there are many naturally occurring carcinogens (cancer-causing compounds) in food that are a normal part of our diet. Ames (1983) has pointed out that plants in nature synthesize toxic chemicals in large amounts, apparently as a primary defense against bacteria, fungi, insects, and predators. The variety of these toxic chemicals is so great that chemists have been identifying and chemically characterizing them for over 100 years. More recently, tests have shown a large number of these natural toxic chemicals to be carcinogenic.

Once again, the evidence supports our recommendation for a wide variety of food with a somewhat decreased intake of calories as the safest diet. In this way one minimizes the intake of any single carcinogen, and since we have a natural resistance to small amounts of carcinogens we can reduce our risk of cancer.

There is no way to reduce risk to zero. Scientists in all of their endeavors should attempt to minimize risk, but it always will be present. We must also be sure that the benefits of any of these endeavors justify the size of the risk associated with them, a concept being used more often in food safety consideration as discussed in Chapter 8.

AMOUNTS OF FOOD ADDITIVES USED

The per capita consumption of food additives has been estimated at about 140 pounds per year, and the kinds of additives that make up this amount are interesting to consider. The most widely used food additives are sweeteners, including cane or beet sugar, corn sweeteners, glucose, fructose, and syrups. We consume an average of about 115 pounds of these per year.

The second most widely used food additive is salt (ordinary table salt), of which we use about 15 pounds per year (this may be decreasing due to concerns about hypertension). Note that the 140 pounds has dwindled to 10 pounds per year if we exclude sweeteners and salt. Following these are about 33 different additives, which account for 9 pounds a year. Of these 33, 18 are used either as leavening agents or to adjust the acidity of food. Yeast, sodium bicarbonate, citric acid, black pepper, and mustard are among the most often used of these 18. Added vitamins and minerals are classified as additives and are also included in the total.

Our list is now down to 1 pound, which is spread over the some 1800 other additives we use. The median level of use of these is about 0.5 mg per additive per year–the weight of one grain of salt per year.

For those interested in particular additives, the food products in which they are used, and the levels at which they are used, more detailed information is available from reports prepared by committees of the National Academy of Sciences and described by Mathews and Stewart (1984).

We hope that this discussion has placed some perspective on food additives. They have important uses in food from a flavor, appearance, preservative, and economic point of view. It is hoped they will always help to provide better food at lower price for more people and minimum risk. If an additive does not fulfill these functions it should not be on the market.

To dramatize our point we would like to reproduce in part an article entitled "Toxic Substances Naturally Present in Food" which discusses the safety of natural compounds versus food additives. This paper is reprinted by permission of the author, D. Richard L. Hall, Vice-President, McCormick and Co., Inc. and the Food and Drug Law Institute, the original publishers, 1973.

There is a basic dichotomy in our Food, Drug and Cosmetic Act respecting its treatment of imitation or synthetic foods as against "natural" ones.

This statutory posture is not peculiar to the United States; many European countries, Germany, for example, go even further in according favored treatment to "natural" foods and food ingredients. Even in its extremes, such a policy must still permit the manufacture and supply of food in an industrialized economy, and this often requires a definition of "natural" as remarkable for its ingenuity as for its comprehensiveness. We play this game, too; the law says in Section 402: A food shall be deemed to be adulterated—(a) (1) If it bears or contains any poisonous or deleterious substance which may render it injurious to health; but in case the substance is not an added substance (i.e., a "natural" one) such food shall not be considered adulterated under this clause if the quantity of such substance in such food does not ordinarily render it injurious to health; . . . An "added poisonous or deleterious substance," however, renders the food adulterated unless the additive is used within the other provisions of the Act, such as the Pesticide, the Food Additive, and the Color Additive Amendments. The regulations issued under these amendments extend, though they do not clarify this distinction between "added" and "not added" or as it is more often said, between "natural" and "synthetic." An example is the listing of synthetic and natural flavorings in CFR 121.1163 and 121.1164 where purification by distillation results in a "natural" product, while crystallization generally produces a "synthetic" one. There is an underlying reason for this. We are attempting to express a higher degree of confidence in the safety of "natural" foods and ingredients than in wholly synthetic ones. It is an attempt to apply the results of human experience in the evaluation of safety. If it lacks precision and consistency, and occasionally even common sense, we should consider the alternatives and what can be done by refinement before we abandon the concept.

There is an alternative, or rather a complex and expensive set of alternatives with which we have become increasingly, but often only superficially, familiar. This is the approach of toxicological investigation, employing primarily animal feeding studies. These are an extremely useful tool, providing valuable insights into the degree and nature of hazards which may be associated with a particular substance. But they have strengths and limitations of both testing and experience. The advantages of human experience in assessing safety are:

1. The experience is gained with the species with which we are concerned—avoiding the problem of interspecific differences.
2. The experience is with the diet composition and within the range of dietary levels normally consumed—avoiding the problems of the consequences of untypical methods of administration and of metabolism by pathways not involved in normal feeding levels.

The limitations of human experience are:

1. Controlled experiments cannot ordinarly be run, although comparative epidemiological studies can sometimes be made.
2. It is not possible to determine the limits of safety by test. Such information ordinarily comes from the study of accidental overconsumption or industrial exposure, or from results in some other application, such as drug use.

The advantages of animal studies are that:

1. Controlled experiments can be run—meaning that you can isolate the use or non-use of a particular substance as the single test variable and determine how the response varies with the dose. In observations on humans, this is usually impossible because of the complexity of both our human genetic makeup and our environment.
2. One can determine the nature and extent of the hazard—and damage—to the test animal with a precision limited only by the skill and equipment of the experimenter, since risk to the animal is of no consequence and the pre- and post-mortem observation may be as extensive as necessary and desired.
3. One can do lifetime, and even multigeneration, studies in animals with a short life span.

These are large advantages, but they are balanced by serious disadvantages:

1. The test animal is not the same as the human animal—not even the same as a miniature human would be. The metabolic pathways may, and often do differ; the susceptibility to damage of the individual organs, or of more generalized bodily functions, will almost certainly differ from those of the human. The susceptibility may be greater or less, and usually in a manner and extent impossible to predict beforehand.

In part, because of these differences, it is customary to apply a safety factor, often 1/100, to "no-effect" levels observed in animals when using these experimental results to estimate safe levels in humans. The result of this is to require in animals doses at least 100 times higher than the functionally effective level intended for human food. In avoiding one trap, we fall into another.

Any substance any animal consumes is either excreted unchanged, or in a few cases, is stored (accumulated) or is modified by the body in some way prior to excretion or storage. This modification, or metabolism, generally takes place by one or a few processes (metabolic pathways) which the organism favors over other paths, presumably because in evolving, these have worked out to the least disadvantage to the organism. As the level of intake of a substance increases, these normal pathways become loaded to capacity, and the organism calls upon other pathways or the substance temporarily accumulates. These other paths will often involve intermediate stages which are more toxic, or mechanisms which place a greater strain on the animal. In any case, they are not necessarily related to the paths, and effects, encountered at lower levels. Thus, the second disadvantage of animal testing is that:

2. To obtain an adverse effect, and to provide an arbitrary but large safety factor, feeding levels must be so high, compared with intended human consumption, that valid analogies very often cannot be made.

The demand that everything be thoroughly tested until safety is proved actually comes from a naive, desperate, and quite unsupportable faith in the extent and certainty of the conclusions which may be drawn from animal tests.

One may well point out that where doubt exists about the applicability of animal data to humans, the decisions should always be made conservatively; and if this is the case, why all the fuss? There are at least two rejoinders to this, one of which is obvious from recent events.

"No effect" in animal studies has every limitation of negative evidence. It simply means that under conditions of that experiment, that experimenter did not find anything. It provides no reason to assume that under some different set of experimental conditions, or with better analytical tools, or a more skilled observer, an effect could not have been found.

We should not ignore another aspect. Not to use a particular

> substance because a more or less thorough investigation showed some significant potential of hazard is not to avoid a danger. It merely exchanges one risk, recently estimated, for another risk which is often unknown. Nowhere is this more apparent than in the attitudes, congealed into regulation, with which we regard "natural" and "synthetic" food ingredients. For like an old tin-type, our food laws, regulations, and company policies present these attitudes as they once were, their rigidity exaggerated, as in a tin-type, by the laborious process of recording them.

Let's consider a reasonably elaborate and attractive dinner menu (Fig. 3.1) and evaluate it regarding each food not as it is treated under the law as natural food, but as it would be treated if manufactured from added ingredients.

Among the toxic substances naturally present in certain foods are some cholinesterase inhibitors of unknown structure. Cholinesterase inhibitors interfere with the transmission of nerve impulses; many potent modern pesticides are based on such activity. These are present in measurable quantity in radishes, carrots, celery, and most particularly in potatoes. In the case of potatoes, the alkaloid solanine (Fig. 3.2) is responsible, and is often present with less than a tenfold safety factor between the normal level and levels that have caused human poisoning. Thus fall the first items from our menu.

A number of food contain glycosides which break down during cooking or digestion to yield hydrogen cyanide. Among those with this disconcerting property are almonds and lima beans. This is no idle concern; lima beans, high in HCN, have been the cause of several serious poisoning outbreaks.

Oxalates and free oxalic acid occur in a number of foods—spinach, cashews, almonds, cocoa, and tea. Our menu is beginning to suffer.

Stimulants occur widely in foods. Nutmeg contains myristicin; tea, coffee, cola, and cocoa contain caffeine. Tea contains theophylline, and cocoa, theobromine. Even more ominously, myristicin is a hallucinogen, and occasionally abused for that purpose. But nutmeg also contains small quantities of safrole, a carcinogen. Unfortunately, we used nutmeg on our spinach, and, of course, the depressant alcohol is not tolerable with a reasonable safety factor, and its hazards are well known.

That alone would rule out the liqueurs, but high intakes of menthol have caused cardiac arrythmia and the glycerine in Cointreau is toxic at only small multiples of normal use.

Goitrogens, substances which promote goiter, are present in many foods. The white turnip contains 1-5-vinyl-2-thiooxazolidone, and cau-

Fig. 3.1. Typical dinner menu.

liflower contains a thiocyanate. It would only take about 22 pounds per day of cauliflower to cause thyroid enlargement, and as careful readers of recent adverse toxicological reports know, this is a wholly inadequate margin of safety. The peach, pear, strawberry, Brussels sprouts, spinach, and carrot have all been shown to demonstrate goitrogenic activity in man.

Pressor amines (Fig. 3.3), which raise the blood pressure, present a real hazard to susceptible persons, and especially to those who are taking drugs such as the tranquilizer, Parnate. Since, by this time, we are

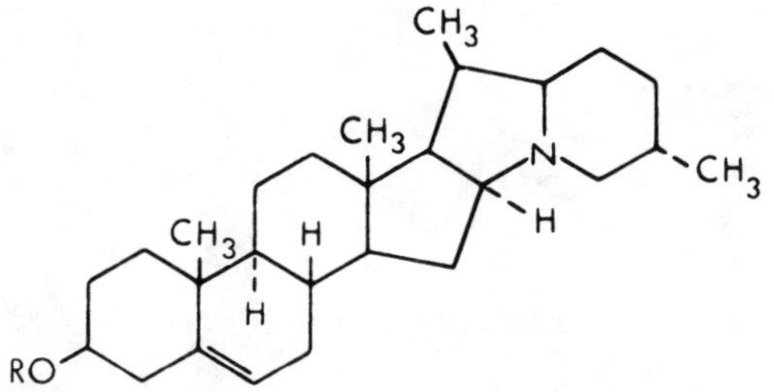

Fig. 3.2. Structural formulas for solanine (R=*l*-rhamnosyl-*d*-galactosyl-*d*-glucasyl) and solanidine (R=H).

sufficiently worried to be gobbling tranquilizers, we should eliminate bananas, pineapples, cheese, especially Camembert cheese, and wine.

Now, we probably need pressor amines, though their essentiality has not been conclusively demonstrated. But this dilemma is presented even more sharply by several of the vitamins—A, D, and K—and several of the essential minerals, which we could not begin to tolerate at 100 times normal consumption levels. But our rule is sacred, and the vitamin D and A in egg yolk and butter, vitamin D in milk, and the zinc and arsenic in seafood rule them out. In addition, egg yolk is reportedly carcinogenic in the diet of mice.

Most of you have heard of the recent concern over the nitrite or nitrate content of foods, and the proved capability of these substances not only to cause methemoglobinemia in man, but also to be transformed in the stomach to the potent carcinogens, the nitrosamines.

These involve not only the cured meats such as ham and bacon, but certain vegetables if they have been fertilized—spinach particularly. Fi-

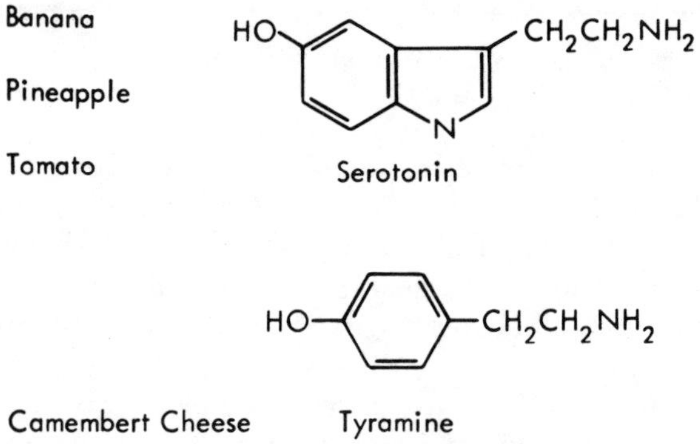

Fig. 3.3. Structural formulas of pressor amines serotonin and tyramine.

nally, smoked foods almost inevitably contain small amounts of the polynuclear aromatic hydrocarbons, and the role of these as dietary carcinogens in man is confirmed by epidemiological surveys of the northern European countries where smoked foods are much consumed, and cancer of the stomach is unusually common.

We can, perhaps, retain the rolls if we can ignore the ricket-promoting factor in yeast and the hazards of amino-acid imbalance. Butter has been eliminated, although, if it were devitaminized, we could retain it, labeling it, of course, for added color.

If space permitted, it would be interesting to speculate on how foods would be labeled, especially if complete declarations of naturally occurring ingredients were required, as in the case of added substances. Some interesting statements would be needed—degoitrogenized, imitation cauliflower, for example.

Hazardous, but essential substances, like the fat-soluble vitamins, could be available on a prescription basis. Those of less clearly justified merit, for example the pressor amines and alcohol, would be available on a non-refillable prescription only.

In all this nonsense, however, there is a serious point. For safety is a serious matter. The whole thrust of this article is that all sources of relevant information should be used. Indeed, this is the underlying concept of general recognition of safety (GRAS), in which both experience based on common use in food, and scientific procedures may be used. Combined, they are still insufficient, and always open to new evidence. Animal testing may be of crucial value—but it may also be irrelevant to human safety. Human experience, for all its directness, may remain an enigma. The utility of both should be improved. In part, this may be achieved by using in animal tests those species previously shown for each instance to be suitable metabolic models for man, instead of those that are handy, cheap, or customary. We need more, and more detailed, national dietary studies coupled with better reporting and analysis of individual health.

Let there be no misunderstanding on one point. We must use the mass of human experience, not mass human experiments. By this, I mean that prior to the broad intentional use of a material in human food, we should have information from animal and human studies which allows expert judgment to conclude with confidence that use of the material will not significantly increase overall risk. But, we must recognize that experience is the final determinant, no matter how encouraging the results from animal tests. We need not only to recognize this, but to improve our utilization of feedback from human experience, in improving our quality of life.

BIBLIOGRAPHY

AMES, B. N. 1983. Dietary carcinogens and anticarcinogens. *Science* 221 (4617), 1256–1262.

BENARDE, M. A. 1975. The Chemical We Eat, p. 108. McGraw-Hill Book Co., New York.

BOFFEY, P. M. 1982. Cancer experts lean toward steady vigilance but less alarm on environment. *The New York Times.* March 2, pp. C1–C2.

COMMITTEE ON DIET, NUTRITION AND CANCER. 1982. Diet, Nutrition, and Cancer. Assembly of Life Sciences, National Research Council. National Academy Press, Washington, DC.

LARKIN, E. T. 1983. Herbs are often more toxic than magical. *FDA Consumer,* Oct. 5–10.

MATHEWS, R. A., AND STEWART, M. R. 1984. Report summarizes data from food additives survey. *Food Technol.* **38**(3), 53–58.

ROBERTS, R. R. 1981. Food Safety, pp. 5, 242. John Wiley & Sons. New York.

SALHOLZ, E., AND SMITH, J. 1984. How to live forever. *Newsweek,* March 26, p. 81.

4

Food Preservation

To one living in New England in the middle of January, the thought occurs of how good a vine-ripened tomato would taste and how long it has been since that particular flavor was savored. There is no disagreement from any scientist or technologist that a fresh vine-ripened tomato tastes better, looks better, and probably is more nutritious than any processed or preserved tomato that one could find. Unfortunately, in January in many parts of America and the world snow, cold weather, and generally unfavorable conditions do not allow for the growth of vine-ripened tomatoes. This is the case with many foods, so in order to have a nourishing, acceptable, safe, and reasonably priced food supply the year round, we must preserve and protect our foods.

Preservation and processing begin when one prepares the soil for planting. In order to obtain maximum yield, as will be discussed in Part II of this book, one must irrigate appropriately, fertilize appropriately, and prepare the soil for the arrival of the plant. The growing plant is subject to insects, rodents, weeds, and other unfavorable conditions that can destroy tremendous amounts of food prior to the time it is harvested. Surveys indicate that in certain areas of Africa, up to 50% of the fish harvested then dried is lost as a result of insect infestation. Therefore, substances such as insecticides, herbicides, and fertilizers must be used in appropriate and judicious amounts.

Assuming that we can harvest a reasonable supply of food materials from our fields, we must determine what to do with these materials in order to keep them acceptable, reasonably priced, and as highly nutritious as possible for a given period of time. Our problem is to prevent the deterioration and spoilage of foods when they are fresh. We have all seen instances of foods going "bad" when left out of the refrigerator or even when left in the refrigerator for too long. Some of this spoilage is due to chemical change, some to biological change, and some, of course, to microbial growth on the food. Many consumers think of spoilage in terms of the growth of microorganisms and the possibility of food poisoning. But this is only one part of the picture. The other extraordinarily important part includes the chemical and biological changes that the food undergoes upon storage. These changes affect flavor, odor, and appearance and may make the food inedible.

Let us look for a moment at the history of food processing. How long has it been going on? Why was it used? Has it been essential? Is it essential now? These are questions we must answer when considering the safety and desirability of food processing.

Food processing has been going on for centuries. Early humans had to rely on growing plants. Animal foods, unfortunately, became inedible fairly quickly because early prehistoric humans did not know how to process or preserve them. In ancient Rome, runners were sent to the Alps to obtain ice with which the aristocracy could keep their foods cold. Spices have been used for centuries, and they were very valuable. For instance, a pound of mace was worth three sheep or half a cow. Pepper, the most valuable spice of all, was measured in terms of human life; it could be used to buy slaves or have men killed. Wars were fought over some of the colonies that praduced spice for such countries as Spain, Portugal, France, Holland, and Great Britain. Why were spices used? In Europe they were used during the Middle Ages to stretch the inadequate supply of food that was available. At that time, most people used spices in order to mask or disguise the flavor of "spoiled" food that was still nutritious but would, because of its flavor, be totally inedible if unspiced. This proves that even long ago people were aware that nutrition was not an isolated phenomenon; food had to be acceptable in order for people to enjoy its nutritive value. Some spices were used because of chemicals they contained. For instance, the cloves used in cooking hams contain a chemical called eugenol, which inhibits the growth of bacteria, a fact known in ancient times. Interestingly, cloves are still used today as a flavor ingredient in cooking hams, but our processing techniques have developed to the point that cloves are no longer required as a means of inhibiting bacteria. Cultural habits have lingered, however, and, today, cloves are used mainly for flavor.

Salt was the most valuable of all the additives. Animals wore paths to salt licks, men followed these paths through the forests and they became trails, then roads and settlements arose. In global terms the original travel routes were known as "salt routes" since salt was one of the principal trading commodities and sixth century Moorish merchants traded salt for gold on an equal basis.

This great value was due to its effect as a preservative in foods and as an antiseptic for wounds. The word salt is derived from the Roman word "*sal*" from Salus, the goddess of health. The Via Salaria (salt road) was one of the busiest roads in Rome, and a Roman soldier's pay, part of which was in the form of salt, was called "*salarium argentum*" from which the word salary is derived.

The sanctity of salt is evident both in the Bible and in the superstitions of the Middle Ages, and its social significance is apparent in that the status of a dinner guest was judged from the relationship between their seating and the silver salt-cellar on the table. Thus we have the expressions "above or below the salt."

Early in history taxes were placed on salt and anyone still unconvinced of its historical importance should look at the Oxford English Dictionary where four pages are devoted to salt.

FOOD SPOILAGE AND DETERIORATION

The type of spoilage that does not endanger the consumer's health should be mentioned, but we should probably spend more time understanding how food is made safe. This is not to imply that chemical and biological spoilage is less important than microbial spoilage; rather, you may be more interested in spoilage that presents a danger to you. In order to clarify these two types of spoilage, it might be simpler to call one deterioration and the other spoilage.

Deterioration of food is commonly seen by the consumer, but not always recognized. Consider how many times you have smelled off-flavored milk or bacon, ham, or other fatty meats that have stayed in the refrigerator too long and have acquired a peculiar odor. That is deterioration. The deterioration, or rancidity, of fatty meats is due to the chemical break-down of fat molecules, forming compounds that have certain odors and flavors. Milk deterioration is caused by microorganisms that break down the protein in the milk to smaller compounds, which have a very definite flavor and odor. However, such microorganisms are not harmful to the consumer and so this is not considered spoilage. Among other types of deterioration are the wilting of lettuce, the toughening of beans and peas, the loss of texture in strawberries, and the drying out of oranges. All of these are changes that you can see as your foods mature and finally become inedible.

When you open a can of food, you never worry about such deterioration as long as you eat the food right away. Yet that food has been in the can much longer than you could hold anything fresh in the refrigerator or in your kitchen. Processing—in this case, canning—has stopped deterioration until the can is opened and the food exposed to oxygen.

Many people feel it is unsafe to leave a food in its can after the can has been opened. This is completely untrue from a microbiological standpoint; the can is probably freer of microorganisms than any dish you might put the food in. During the heat process to which it is exposed, the can has been heated to temperatures far higher than those produced by any dishwasher or those used in hand-washing dishes. However, some contamination might occur from the interior of the can and cause problems such as off-flavors in certain acid products. Therefore, this practice should be avoided. Deterioration is at times prevented by packaging. Examine a package of potato chips. The container probably suggests that the chips not be exposed to sunlight; quite often, the container itself is colored or opaque. This is because the fat in the potato chips, when exposed to sunlight, becomes rancid and forms the odoriferous compounds referred to previously. There is a reason for almost every package or process that is used in modern food processing. Natural flour or cereals must be stored in the refrigerator and cannot be exposed to room temperature for long periods of time. If they are so exposed, fat rancidity occurs and they become inedible. If grains were not milled to remove most of the fat, then the flour that is formed from such grains could not be stored in large bins at room temperature as flour may be stored today. This would seriously affect the amount of flour being shipped to developing countries with hot climates, because of the rancidity that might occur.

Therefore, there are technological and scientific reasons for milling grains.

Spoilage, the term we are applying to microbiological hazards, cannot be ignored by the consumer. We really do not know just how prevalent the hazards from food-contaminated bacteria and other organisms are. The reasons are that the illnesses resulting from such food contamination are not always evaluated by the victims or reported by physicians. Nevertheless, informed estimates indicate that perhaps as many as 20 million Americans suffer the effects of bacteria-contaminated food each year. This does not seem to be an especially great number, but such illnesses are occasionally fatal, and only the common cold causes more loss of work. The contamination is due almost entirely to unsanitary handling practices in restaurants, catering establishments, and, in the home. Stomach upsets or worse used to occur so frequently at church picnics, as a result of home handling of foods, that "church picnic" nearly entered the medical lexicon as a descriptive term. It should be stressed that the foods received in the home today are rarely, if ever, the cause of microbial hazards. Only after they are opened and handled do they become hazardous due to microbial growth.

Types of Food Poisonings

There are three major types of microorganisms that affect foods—bacteria, yeast, and molds. There are about 50 different species of microorganisms found in food that can cause deterioration. Bacteria are the major health problem. Yeast are not usually considered poisonous, but they do cause food to decay. A few toxins that are formed from some molds found in foods are dangerous and can cause a variety of health problems.

We are all familiar with ptomaine poisoning, which results from eating certain foods that have been contaminated by bacteria. However, the term ptomaine poison was developed before we understood bacterial poisons. The term does not mean anything and should not be used.

There are two types of food poisoning caused by bacteria. The first type, food infection, is caused by the presence within food of bacteria that can grow rapidly within the human intestinal tract, producing symptoms such as diarrhea and vomiting. The organisms that cause this type of food poisoning are *Salmonella* and *Shigella*. The second type of bacteria-caused food poisoning is food intoxication. As the name implies, food intoxications are due to toxins that are produced by bacteria and then ingested by a human. In this case, it is not the bacteria themselves that cause the food poisoning but the toxins produced by the bacteria in a food. *Clostridium botulinum, C. perfringens*, some molds, and *Staphylococcus aureus* produce harmful toxins.

In most cases of food poisoning, whether it be food intoxication or food infection, the symptoms include diarrhea, nausea, vomiting, and related symptoms. However, *Clostridium botulinum* causes death in about 60% of all cases. Because of the dehydration and weakening that they cause, other varieties of food poisoning can result in debilitation and death if they atack a young baby or an elderly person. Food poisoning, although not common, is a dangerous and of major concern to the modern food-processing industry.If a food has a terrible off-odor or off-flavor, the consumer will not accept it and thus, of course, will not obtain any nutrients from it. However, a food that does not exhibit off-flavors or off-odors may still be subject to microbial contamination and, if consumed, may result in sickness and perhaps death. Interestingly, few cases of food poisoning can be blamed on processing or technology; most are due to preparation and handling after the food is purchased.

However, we must be vigilant since there have been recent occurrences of botulism poisoning in restaurants; the unlikely source was onions in one case and potatoes wrapped in foil in another case.

Other pathogens such as *Listeria monocytogenes* and *Yersinia entero-*

colitica have been found to grow in milk at refrigerator temperatures, an unexpected occurrence. Quality control, preservation, and processing are indeed essential.

For the consumer, the handling of food should always be viewed in terms of proper sanitation. Even people who do their own dishes often exhibit extremely unsanitary practices in the use of their dish towels. How many times in a home have you seen someone wipe some dirt off the floor, scrub a child's face, or even wipe a nose with a dish towel and then blithely go on drying the dishes with the same towel? Certainly, the contamination from the floor, the face, and the nose could add to the microbial load on the dishes. This contamination could later be transferred to food on a dish and cause food poisoning. In many households wooden meat boards may be contaminated by *Salmonella* or some other organism. The organisms get between the fibers in the wooden board and may later be transferred to another food, even after it is cooked, which is then ingested by the human. We are seeing a resurgence in the use of wooden cutting boards in the kitchen and the consumer should be forewarned as to its potential dangers.

Staphylococcus causes a food intoxication that is a very common form of food poisoning. It is the one that hits us very shortly after eating; incubation time is only 2 to 4 hours after ingestion. Violent diarrhea and vomiting are the symptoms. *Staphylococcus* contamination is commonly found in cream pastries, tuna, canned chicken salads, and similar mixtures of foods. It is extremely important to refrigerate such foods immediately after handling, because the bacteria cannot grow at refrigeration temperatures and the toxin cannot be produced. Heat will not destroy the toxin that *Staphylococcus* produces. Therefore, both sanitation and refrigeration are needed in order to prevent the formation of the *Staphylococcus* toxin. Contrary to popular opinion, mayonnaise and salad dressing, when added to salads or sandwiches actually retard the spoilage and growth of harmful microorganisms. This is due to the fact that they contain acetic acid (vinegar), which increases the acidity of the product and retards bacterial growth, as will be discussed later in this chapter.

Clostridium perfringens is usually contracted from eating contaminated meats, eggs, gravies, and other protein foods. The organism is a vegetative cell within a protective coating called a spore. When ingested, spores may grow inside the intestines and produce a toxin, which in turn causes the symptoms of food poisoning—vomiting and diarrhea, but not death—in the human.

Because it is so rare and yet so devastating, death from botulism poisoning is a sure way to make the newspapers. *Clostridium botulinum* is an anaerobic organism, that is, it cannot grow in the presence of oxygen.

It is normally found in canned foods and in sausage meats, where oxygen is excluded. Like *Clostridium perfringens*, it has a vegetative cell and a spore. But unlike the perfringens spore has to vegetate within the food material in the absence of oxygen and produce a toxin prior to ingestion. It is the toxin that causes the poisoning and often results in death. Adequate processing of foods—in particular, heat processing—will destroy the vegetative cells and the spores of *Clostridium botulinum*. Misreading the pressure cooker gauge, not processing the food long enough at a high enough temperature to insure the destruction of botulinum spores, or faulty equipment is the reason for the larger number of botulism outbreaks in home-canned food than occurs in industrially processed food. Approximately 800 billion cans of food were produced before 1972, and during that time only five deaths from *commercially* canned foods were reported. In 1973, six deaths from *Clostridium botulinum* were reported, but only one of these was caused by industrially processed food. Several other outbreaks of food poisoning from *Clostridium botulinum* have occurred since 1973, particularly in the canned mushroom industry, but this problem is under control and is of little consequence today.

FOOD PROCESSING

Having mentioned some of the microorganisms that cause spoilage and some of the deteriorations that take place in foods that are not processed, it would now be appropriate to discuss some of the methods whereby foods are processed. Processing is simply a way to maintain safety, acceptability, and nutrition in our food supply for an extended period of time.

How do we prevent the growth of bacteria in a food material? This is probably the fundamental question in food processing, because the safety of food is of the utmost importance. Once we have established the appropriate process to insure a safe food, we may then consider ways to prevent deterioration and ways to make the food more appealing in flavor, color, texture, appearance, and so on.

Microorganisms or bacteria are like you and I: they are born, they have a life span in which they must eat food and excrete waste, and they die. The question is, how do we control the life, growth, and death of bacteria? There are several possible ways of doing this.

- Remove the food source
- Change the temperature
- Remove water
- Change the environment

Remove the Food Source

This method is not a viable one in that taking away the food source of the bacteria is, of course taking away the food source for us as well. Therefore, this alternative can be dispensed with immediately.

Change the Temperature

Our second option is an extraordinarily effective way to kill bacteria and/or prevent their growth, thereby preventing both deterioration and microbial spoilage from occurring. The temperature can be changed in a number of ways. Refrigerators keep fresh produce and opened processed foods for a longer period of time than such food would keep at room temperature. The refrigerator's lower temperature slows the growth of bacteria so that they cannot multiply to such an extent that they cause food intoxication or food infection. The refrigerator also prevents deterioration such as the wilting of lettuce, the rancidity of fats, and the souring of milk. A refrigerator adjusted to 40°F can just about quadruple the keeping time of most foods.

Freezing is another method of preservation based on temperature control. Freezing prevents the growth of microorganisms; it does not kill them. Growth is prevented in two ways. (1) It freezes the available water so that the microorganisms do not have enough water to grow. (2) It lowers the temperature to a point where the microorganisms either cannot grow or grow very slowly. Home freezing has become a very common way to process food, but although it produces a very desirable product, it is uncommon to find home-frozen food of as good quality as industrially frozen food. This is not to cast aspersions on the freezing capabilities of the average consumer or on the average consumer's kitchen. One of the major dimensions of processing a high-quality food product is the length of time the process takes. That is, the quicker you can freeze the food the higher in quality it will be. Less tissue damage will occur, and, therefore, the texture will be better and flavor loss, color loss, and nutrient loss will all be lessened. Industrial processes use such techniques as freezing by liquid nitrogen (−280°F) and blast freezing—that is, using a fan along with the freezer to create a wind effect. Most of us have heard about the wind-chill factor that is computed for areas with cold winters. A low temperature coupled with a strong wind can result in a wind-chill factor of, say, −60°F or colder. Industrial blast freezers attain the equivalent of very low temperatures and quick-freeze the food. We cannot do this in the home. The normal home freezer compartment takes longer to freeze food than the industrial freezers.

We should add that it is necessary to blanch a food prior to freezing it. Blanching simply involves subjecting the food to a high temperature (approximately 190°F) for a minute or so prior to freezing. This process destroys the enzymes present in vegetables. Remember from Chapter 2 that an enzyme is a biological catalyst; that is, it hastens reactions. If the enzymes are not destroyed by heat, certain chemical reactions will occur while the food is frozen, and off-odors and off-flavors will result. These chemical reactions constitute a deterioration rather than a spoilage: they do not present a health hazard, but they create a real problem in quality.

As a general rule, if you thaw food, do not refreeze it. If, however, you thaw food slightly in the refrigerator, it is quite safe to refreeze it. Problems arise when one leaves food at room temperature for an extended period of time; the food thaws and then stays at a temperature at which the microorganisms (which were not destroyed by freezing but merely rendered dormant) begin to multiply. The organisms may or may not produce toxins, or they may multiply to the point where they are likely to cause a food infection. If the food is refrozen, the toxins that were produced or the bacteria that were not destroyed may cause food poisoning when the food is ingested at a later time. A good rule of thumb is that it is safe to refreeze a food if ice crystals are still present in it.

The other obvious way of utilizing temperature to control or kill microorganisms is heating. Cooking is the most obvious example of using heat to destroy microorganisms. We do not normally use cooking as a method of destroying microorganisms; we use it as a method of making food more palatable. However, cooking does destroy microorganisms and heat is the major method used in food processing and preservation by canning. This method was originated by a Frenchman named Nicholas Appert during the time of Napoleon. Napoleon needed foods that could be stored for use by his armies, and he offered some 12,000 francs to anyone who could discover a way of producing such food. Appert invented a method of sealing foods inside a container and then heating it. He thought that the foods were made safe by the destructive action of heat on the bacteria in the container. Home canning is becoming increasingly popular but can be dangerous. The local land-grant state college or university and their extension service or the many legitimate books on the subject should be consulted to find out the exact time and temperature that apply to a given food in a given size of container. In the food-processing industry, mathematical analyses are used to predict the specific time and temperature necessary to destroy hazardous microorganisms for any size of container and any type of food. Unfortunately, accidents do occur on occasion, but, they are extraordinarily rare. The higher the temperature used, the shorter the heating time required. The modern food-processing industry has been able to achieve higher and higher temperatures and shorter and shorter times. This combination manages to kill the bacteria effectively but does not destroy as many nutrients.

Remove Water

The removal of water is the third way in which we might prevent the growth of microorganisms. Bacteria require, just as you and I do, a certain amount of moisture to grow and survive. The drying process simply removes enough moisture from the food so that the bacteria cannot grow, multiply, and/or produce toxins. Drying is probably one of the oldest methods of food processing. Sun drying, which is used to make raisins from grapes, is a very old method and one of the fundamental ways of drying a food material. The American Indians used to make pemmican, a meat and berry mixture that was pounded together and then dried in the sun. The berries contained sugar and salt, which aided in binding the water so that microorganisms could not grow, and the acids from the berries also helped prevent microbial growth during the drying process. At the time pemmican was being made these factors were certainly not known, but the Indians found that if certain methods were not used in the preparation of this food, people who ate it became sick. As with freezing and heating, the quicker the drying the more nutrients retained. It has become a challenge to modern technology, therefore, to develop methods to speed up the drying process. In the past few decades, drying has become a finely tuned technology that produces very high-quality, safe foods. From ancient methods of sun drying there evolved other methods, such as tunnel drying, whereby a food is put on a belt moving through a tunnel where a hot-air draft moves in the direction opposite the travel of the food. This decreases the time of drying by a tremendous factor. Spray drying was developed as a means of improving quality further, and this is how most dried milk is produced. In spray drying, fluid milk is fed into an atomizer, which breaks the milk down into fine particles and sends them into a heated chamber. Each tiny particle is then dried very quickly, and the resulting power is collected as dried milk powder as we know it today.

In the freeze drying process a food is frozen and then some heat is applied under reduced pressure (vacuum) in order to sublime the ice into water vapor; that is, the ice is transformed directly to vapor without going through the water phase. By having the ice change directly to vapor, the cellular tissue is not broken down as much and the food maintains more of its physical integrity. This means that when the water is returned, the freeze-dried food is more similar to the fresh food than most dried foods would be. Unfortunately, the cost of freeze drying still prohibits the marketing of a complete array of freeze-dried foods.

The amount of water that has to be removed from a food in the process of drying is not as high as one might think. What is important is not the amount of water present as much as the availability of the water within the food. This "availability" is known as the water activity of a food and has to be below a certain level in order to prevent the formation and growth of microorganisms. For instance, the semi-moist dog foods that are currently on the market today do not feel or look like the completely dry product that we are used to seeing. This is because not all the water has been removed from these foods. Rather, sugar and salt have been added, which binds the available water so that it cannot be utilized for bacterial growth. This "binding effect" is achieved by the use of many food additives, which, by lowering the water activity, produce a very acceptable, safe, and nutritious food. Again, we have a trade-off between nutrition and producing a safe, long-lasting food. If we do not remove all the water, we probably will not destroy as many nutrients. By using food additives such as sugar and salt or sorbitol and mannitol (simple sugars), we can bind some of the water that is available and thereby prevent it from being used for bacterial growth.

Change the Environment

Irradiation of food is another method of preservation whose role in the U.S. food system may soon be expanding. The decision of the Environmental Protection Agency (EPA) to suspend the use of ethylene dibromide (EDB) as a fumigant for insects in stored grains, machinery, citrus fruits, and papaya has created an immediate need for an alternate method, which irradiation might fill.

Food irradiation involves the use of energy in the form of ionizing radiation to destroy insects or microorganisms without changing the food as heating does. Theoretically, an orange can be irradiated to destroy any insects that could cause spoilage without changing the orange. Irradiation at the levels required in food does not cause the food to become radioactive or dangerous in other ways. It is simply an alternate use of energy to preserve food.

Currently, 21 countries allow one or more of 31 foods to be irradiated. In the United States, the only irradiated foods available to consumers are some spices and powdered vegetable seasonings approved by the FDA in July 1983. However, in February 14, 1984 the FDA issued a proposal in the *Federal Register* that would allow low-dose irradiation to delay ripening of fresh fruits and vegetables and to kill insects that infest food. Further, many scientists are proposing its use for poultry and shellfish, which would increase their storage life. Another use would be to treat pork to reduce the risk of trichinosis. Still a problem today, pigs can become infected with *Trichinella spiralis,* a parasite, if they are fed garbage. The parasite becomes lodged in human muscle if it is not destroyed. Current control involves heating the entire piece of meat (ham, pork, etc.) until an internal center temperature of 137°F is reached.

Irradiation has the potential to play an important role in our food supply system, and its judicious use will be beneficial.

Just as we find certain environments harmful to our growth, such as various forms of pollution, bacteria also find certain environments harmful. Environmental factors have been manipulated for hundreds of years as a means of controlling the quantity of microorganisms in a food material. For instance, fermentation is a microbial process whereby microorganisms grow and produce an environment that is unsuitable to the growth of other microorganisms. This is how beer, hard liquor, and sauerkraut are made. Fermentation gives nonharmful microorganisms an environment in which they can thrive. These microorganisms grow and excrete acid that inhibits the growth of harmful microorganisms. Another means of producing an unhealthy environment for certain microorganisms is changing the acidity of the food product. Acidity is measured in terms of pH. The lower the pH the higher the acidity, and the higher the pH the lower the acidity. *Clostridium botulinum* will not grow in a food whose pH value is less than 4.5. Therefore, acids are added to certain food products in order to decrease the pH to below 4.5, insuring that *Clostridium botulinum* will not grow. These acids are food additives, and quite often they occur naturally: malic acid comes from apples, citric acid from oranges, and acetic acid from vinegar. Many food additives are simply natural compounds added to a food material to prevent the undesirable formation or growth of microorganisms and the subsequent changes or harm they might cause.

Smoking a food is another method of changing the environment so that it inhibits the formation or growth of microorganisms. Smoking is as old as any of the other processes previously mentioned. It has been used for hundreds of years as a means of producing meat that is long-lasting, nutritious, and acceptable. The smoke dries the surface of the meat and thereby prevents the growth of microorganisms to some extent. In addition, some of the compounds produced in the smoke inhibit the growth of particular microorganisms. However, the smoking of meat produces compounds that may be carcinogenic, indicating that there is a certain risk as well as a certain benefit to enjoying certain foods. Each person must weigh the benefit and the risk involved and then make the choice. If the risk is minimal and the benefits great then we will continue to enjoy these foods.

Filtration is another method whereby certain microbes are removed from food materials. In some cases, pasteurization of beer is accomplished by filtering so that the microbes and in particular the yeast are left in the filter and the beer passes through. As a result, the beer does not have to be heated prior to canning. This type of filtration has opened the way for several brands of "real draft beer" in cans.

Another method of manipulating the environment is, of course, packaging. Any home canner knows that the way to prevent mold formation on jams and jellies is to pour some hot wax over the top of the preserve. When cool and solidified, this wax prevents the oxygen in the air from coming into contact with the preserve and thereby prevents the formation and growth of molds. Vacuum packing accomplishes the same purpose, removing the oxygen by suction or by some other method prior to packaging. Vacuum packing is simply modern technology's way of placing a wax coating on today's foods.

We hope that this chapter has given the reader an understanding and "feel" for the techniques used in food processing. We also hope that after reading this chapter you will realize that processing is not magic, is not bad, is not good; it is just a means of preventing the spoilage and deterioration of foods. We must process foods because we do not live in a world in which we can grow foods any time we want to. Therefore, we must find a suitable means of providing fairly inexpensive, safe, acceptable, and nutritious foods the year around. Preservation provides us with this means. Preservation and technology along with advances in cooking appliances such as microwave ovens have also provided the consumer with great convenience. Prepackaged foods that are ready to heat and serve have liberated our society to engage in activities that they feel are more meaningful than household labor. Convenience foods have in some respects provided us with a kind of domestic help. Not everyone wants to use convenience foods and not everyone should. However, there are a wide variety of convenience foods available today, admittedly at a fairly high price, that consumers may utilize if they so desire.

Food processing is necessary in today's world to provide the quality of life and variety of foods essential for good health. It is hoped that the production of safe, high-quality and nutritious foods will continue so that more people are fed better throughout the world.

5

The Road to Fitness and Health

A discussion of fitness and health and its relationship to food necessarily involves chasing a moving target since the issues involved cannot claim science for their sole ancestry. In part, this is because eating is a social act. This aspect of food was succinctly described by Farb and Armelagos (1980) who stated that "All animals feed but humans alone eat." Humans also like and dislike individual foods for other than sociological reasons. Fallon and Rozin (1983) have suggested four major categories for food rejection, each with its own profile of psychological characteristics: (1) Distaste, a sensory rejection with no objection to the presence of such foods in the body or in untastable amounts of other food. (2) Disgust, based on what the substance is with an objection to having it in the body or in other foods. (3) Danger, based on anticipated harmful consequences of eating the food. (4) Inappropriate, these substances are not considered to be food.

Image is another important aspect in food choice, and Hertzler *et al.* (1983) have given examples of some image foods: supercultural foods or prestige foods (snobbish, exotic, gourmet), body image food (beautiful, athletic), reward foods (good behavior, success), status foods (rich, elite), magic foods (cure-all, quick weight loss), and foods for sickness.

In addition to all these critical effectors of food choice, Clydesdale (1979) has pointed out that purchase appeal must also be considered within the framework of such factors as nutrition, cost, safety, availability, convenience, and quality. Indeed, it becomes obvious that the consumer is faced with many questions when making food choices: What foods should I eat? What foods can I afford? What foods do I want to eat? What foods should I eat to show people who I am?

These questions, along with a barrage of alternative advice offered by government, academics, industry, consumer groups, self-appointed nutritionists, exercise gurus, neighbors, and myriad other sources often result in confusion and frustration.

We are beseiged by print and electronic advertisements of muscular young men and women using exercise machines that would fit into a Star Wars production, making us feel guilty about the fact that we do not look like they do, but never telling us that it may be genetically impossible for us to look like that. The claims of many of these ads stand in sharp contrast to the modest Statement on Exercise published by the American Heart Association (1981) and most of the scientific literature.

This enthusiasm for health through technology has replaced much of the natural movement of the 1960s and 1970s but has brought with it another set of problems. Both beliefs demand a simple answer to the complex questions of health. Some of this "high tech" fitness has resulted in an increased use of drugs in sports, and, unfortunately, "athletes rely as much on rumor as on conflicting scientific reports about the effectiveness and danger of drugs to improve performance" (Zurer, 1984). In spite of the great potential risks involved, many athletes feel that it is worth it to get the competitive edge. However, due to increased scrutiny some athletes are avoiding detectable drugs and are turning to such sophisticated scientific procedures as blood doping, which uses the athletes own blood rather than chemical substances to boost performance. Several weeks before competition, blood is removed from an athlete and stored. Just before a meet, when the competitor's hemoglobin level has again reached normal levels, the stored blood is reinfused, increasing the concentration of hemoglobin in the athlete's system and potentially improving oxygen capacity and endurance. However, the efficacy, legality, and morality of this technique are in question.

The fitness movement has effected not only the young, but people of all ages. More importantly, aging has become an accepted phenomenon, and thankfully the long fruitless search for perpetual youth seems over. This is due in part to both changing demographics and recognition of the simple fact that age is not a drawback in achievement levels in many areas of society.

A healthy image and the notion of being responsible for one's own health is rather refreshing. However, food is being blamed or credited for nearly everything, and the hypochondria sweeping the country is forcing unwise decisions on naive and gullible consumers. There is no question that what you eat affects how you feel and is thereby an integral part of your health. Food, however, will not provide cure-alls for anyone. An appropriate diet allows people to live up to their genetic potential and feel a sense of completeness about their body. However a person 5 feet tall will not grow suddenly to 7 feet and become a basketball star by eating special foods. It is virtually impossible to discuss each and every health recommendation and its merits or demerits, but perhaps a look at the more common ones may be helpful.

EXERCISE

It is very difficult to find in the scientific literature a definitive study on what exercise will and will not do for you. Certainly it can be said that genetics are extremely important, and most of us will never look like Ms. or Mr. America no matter what we do. Those rippling muscles, thin thighs, and flat stomachs are in part inherited, and although we can improve on what we have, we cannot make a wooden building from bricks. Forbes (1982) studied the effects of exercise (jogging, weight lifting, exercise machines, etc.) on the total amount of lean body mass in athletes and concluded that ". . . exercise and/or training does not result in a significant augmentation in lean body mass for the majority of subjects thus far reported." He also points out that although athletes tend to possess a larger lean body mass than sedentary individuals, this is not proof that training accounts for the difference. The athlete may have possessed a larger lean body mass from an early age. Unquestionably, training decreases the weight of an individual because of the calories expended, as shown on Table 6.4 for selected activities, but it is questionable if you can increase lean body mass over your genetic predisposition.

This should not in any way detract from the positive effects of exercise. Weight control is certainly a benefit due to energy expenditure as well as the possibility that aerobic exercise, such as walking, swimming, jogging, etc., done on a routine basis (four to five times a week) may maintain the basal metabolic rate (BMR) even while you lose weight. The BMR is a measure of the calories required to sustain life in the resting state. If it can be maintained, rather than decreased, during weight loss the amount of food does not have to be continually reduced. Women, in particular, because of size, become very frustrated because after a certain weight loss, they reach a plateau where they virtually cannot lose any more weight on a nutritious diet. In such cases exercise is essential to maintain health and lose weight.

Oxygen capacity is generally improved through aerobic exercise, and the resting pulse rate often declines. Exercise is also thought to decrease calcium loss from bone and thereby decrease the risk of osteoporosis and bone fractures. Exercise seems to be a factor in controlling blood pressure and in maintaining levels of HDL (the right kind of cholesterol) in the blood, as was discussed in Chapter 1.

It still must be realized, however, that evidence equating exercise with increased life span is contradictory. What can be said about exercise? It is not a panacea, it will not insure protection against disease. However, in moderation it will certainly not hurt anyone, and the evidence suggests that it will help. Don't overdo it—but do it.

VEGETARIANISM

Prior to discussing the pros and cons of vegetarianism we must define our terms. Strictly speaking vegetarianism means a diet completely devoid of all animal products and by-products which means no meat, milk, eggs, cheese, poultry, fish, yogurt, etc. This diet may be dangerous and groups, such as the "vegans" in Britain, who follow such a diet have a long history of poor health. It was discussed in Chapter 1 that such a diet may be deficient in such nutrients as vitamin B_{12}, iron, and zinc among others.

However, most people are ovo-vegetarians (ovo=eggs), lacto-vegetarians (lacto=milk), ovo-lacto-vegetarians, fish-ovo-lacto-vegetarians, or poultry-fish-ovo-lacto-vegetarians. Of course, the more food groups you add to the vegetarian diet the less the risk of nutrient deficiency. Please note, we have recommended a shift in diet to plant foods but not the elimination of meat from the diet. Certainly we should not eat a pound of meat a day but 3–5 ounces of lean meat, fish, or poultry per day contributes to a balanced diet.

ZEN MACROBIOTIC DIET

The zen macrobiotic diet is not part of the Zen philosophy or religion, but rather was established by a person named George Ohsawa. It is a diet that begins with a fairly broad group of food materials, from which certain foods are slowly excluded until one is consuming only brown cereal grains and a sparing amount of water. From Chapter 1, you should realize that this diet can be extraordinarily dangerous to one's health. Mr. Ohsawa once stated that "a zen macrobiotic person cannot be killed by an atomic bomb." This may or may not be true, but a person on this diet may be killed by a deficiency disease.

It has been found that children who were fed a zen macrobiotic diet by their overzealous parents suffered critical nutritional deficiencies. *Newsweek,* in its issue of September 18, 1972, reported that two such children were tested for bone development. Normally, a child's "bone age" should be within two months of his actual age, but in tests a two-and-a-half-year-old child showed a bone age of six months and a one-and-a-half-year-old registered three months. Diagnosis showed that both little boys were suffering from severe cases of rickets and scurvy—and that they had been living for months on the zen macrobiotic diet. Neither child could walk or crawl; the older child had a vocabulary of two words and weighed sixteen pounds; the younger just reached eleven pounds; their hair was coarse and brittle and they were extremely cranky. Ironically, the parents of these youngsters brought them to the pediatrician not because they were concerned about the state of their health but to show the doctor how well the zen macrobiotic diet worked. Concerned friends had tricked the parents into visiting the clinic with their children by telling them that the doctor was interested in zen macrobiotics. The parents refused to abandon macrobiotics altogether, but they finally agreed to give their sons oranges, meat, fish, eggs, and dairy products. After six months the children no longer showed the painful, fragile bones of rickets or the skin eruptions of scurvy, and the eldest had started to walk. Even so, the future for these youngsters is not promising. What is unfortunate about this incident is that the parents were looking for magic from food and ignored the health of their children. Through their own ignorance they have perhaps caused permanent damage.

We do not consider the cultural and historical tradition of vegetarianism to be faddism, nor do we believe that any of the ethnic or religious dietary concepts that are held by many groups constitute faddism. These traditions have been tested down through the ages, and through education and practical experience healthy diets have developed and become a part of the tradition.

ORGANIC AND NATURAL FOODS

The Federal Trade Commission has been unable to come up with a suitable definition of "organic" and "natural." We do know that they always cost more. Do not confuse fresh foods with organic or natural foods. Fresh foods in season are foods of choice and do not cost more nor do they claim magical powers.

CHEMOPHOBIA

Chemophobia (a fear of chemicals) has developed in our nation with some basis. Problems do, and have, existed with pollution due to the misuse and abuse of chemicals in our society. Unfortunately, this has led to a wide scale belief by some that any processed food is bad, that any food manipulated by man is bad, and that the addition of any chemical to a food is wrong. It should be evident from the previous chapters that foods are simply chemicals, humans are simply chemicals, and the only really scientific reason for eating is to replace the chemicals in the human body with chemicals from food. One can only conclude, therefore, that the addition of chemicals to foods, as long as such chemicals are safe, is at times necessary. Unfortunately, however, scientists have sometimes proclaimed on the basis of very poor evidence that certain foods are bad for health and that the consumer should therefore not eat them. Generally, such beliefs are held by a minority of the scientific community. Benjamin Feingold (1975) wrote a book in which he stated that food colorants and many food additives are responsible for hyperactivity in children. This idea was accepted immediately by many even though the scientific evidence was scant. Since that time there have been many scientific studies including one by Rapoport of the National Institute of Mental Health (Kolata, 1982) that failed to find a significantly high correlation between food intake and hyperkinesis. In fact, at a conference on food and behavior in 1982 at the Massachusetts Institute of Technology, it was reported that although folk wisdom says that refined sugars and carbohydrates cause children to be hyperactive, scientific studies show that the more likely effect is to make people sleepy due to their role in the formation of serotonin, a neurotransmitter in the brain which induces sleep (Clydesdale, 1984). The danger of Feingold's diet is that he also proposed excluding 21 fruits and vegetables from the diet, which could induce nutritional risk. Unfortunately, this kind of thinking could lead to an individual attempting to get his nutrients from questionable sources.

Here is an interesting list prepared by Frank Konishi, head of the Department of Food and Nutrition, Southern Illinois University at Carbondale. It describes how much of certain foods must be eaten to obtain the RDA.

Nutrient	Food to Obtain Daily Requirement
Protein	30 slices of bread, 82 "protein pills," or 40 servings of gelatine dessert
Vitamin A	22 pounds of sunflower seeds
Vitamin E	60 raw eggs
Vitamin C	10 apples
Niacin	36 glasses of beer
Riboflavin	230 tablespoons of honey and vinegar
Thiamin	25 tablespoons of blackstrap molasses
Calcium	68 raw oysters
Iron	45 doughnuts
Iodine	6½ quarts of seawater

Obviously, none of the foods on this list are good sources of the nutrients mentioned, although some people believe them to be. The point of reproducing this list is that there is no single source, natural or not, that can provide us with all the key nutrients.

PREVENTION OF HEART DISEASE BY DIET

There is great controversy surrounding the role of diet and heart disease, as was discussed in Chapter 1. Our recommendations for the 80% of the population without an inherited trait for high blood lipids indicates that the exclusion of any one food group from the diet is unnecessary and in fact might be harmful. However, there should be an attempt to lower calories, decrease consumption of meat to about 3–5 ounces of lean meat per day, eat dairy products in moderation, increase consumption of grains, fruits, and vegetables, and use unsaturated fats and oils for addition to foods and for cooking since saturated fats already occur in meats and dairy products.

Heart disease is multifactorial, being dependent on such factors as genetics, smoking, weight, alcohol, stress, etc. Diet can be a part of a program to decrease risk, but it cannot insure prevention.

DIET AND CANCER

Cancer is the second leading cause of death in the United States and perhaps one of the most feared of the life threatening diseases. Cancer is the formation of abnormal cells that grow in an uncontrollable fashion and crowd out other tissues. The exact reasons why this occurs are unknown, but some cancers have been linked to viruses and others are thought to be caused by some agent or agents in the environment. There is also the distinct possibility that some people are more susceptible than others due to genetics.

Since we do not know the exact reasons why cancer occurs, it is understandable that people search for compounds that might prevent or protect their cells from change and conversely avoid any materials that might cause the change.

Unfortunately, in the realm of diet we do not have cause and effect relationships between food and cancer and therefore cannot promise prevention by changing the diet.

However, the committee on Diet, Nutrition and Cancer (1982) has issued a report summarizing the evidence relating dietary patterns with incidence of cancer (see Chapter 12).

It is interesting to note that the recommendations are very similar to what we have been proposing throughout this book, as well as what has been proposed in the USDA guidelines and the McGovern Report which are also discussed in Chapter 12.

The difference between our recommendations and the others is more in how they are stated than in the message they contain. For instance, we recommended the intake of the RDA for all nutrients, which will make you healthier all around. Note in the recommendations (Chapter 12) from the Committee on Diet and Cancer that excessive amounts of nutrients are discouraged—we concur. We also believe that the risk from smoking and alcohol abuse is so much greater than that from smoked meats or fat that they should not even be mentioned in the same list. Smoking is the leading cause of cancer. Accidents, many of which are alcohol related, are the leading cause of death among young people. In fact, the death rate is rising for the 40 million Americans of ages 15 to 24, as a result of accidents and violence often due to alcohol and drug abuse. Therefore, it seem hypocritical and maybe pointless to smoke and drink heavily and still worry about your diet.

As far as other dietary factors are concerned, the committee in response to questions, stated that there was insufficient evidence to reach conclusions on the effects of dietary fiber, cholesterol, vitamin E, chlorinated water, sugar, eggs, caffeine, coffee, or tea. The committee did not examine the effects of "nutritional therapy" or smoking with regard to cancer.

DIETARY CURE-ALLS

It is common for people to search for dietary cure-alls other than those related to heart disease and cancer.

Some of us take vitamin C to cure the common cold, in spite of recent tests that have shown that vitamin C neither cures the common cold nor prevents its onset. There are some indications that the length of time one has a cold may be lessened, though almost imperceptibly, due to the intake of vitamin C.

How many people take vitamin E to increase sexual potency (a completely unfounded belief)? Vitamin E has also been claimed to reduce heart disease and atherosclerosis, a claim that has very little scientific backing.

The list of such fads goes on and on. It is most unfortunate that people are being bilked into taking extraordinarily large dosages of vitamins or other nutrients in order to prevent or cure certain organic disorders that by and large are not curable by the ingestion of food. Although these fads seem illogical, they are followed by many people. Perhaps the reason for this is that it is easy to ingest a food in an attempt to cure a disease; one thinks, "It's worth a try." Unfortunately, this kind of cure-all can be extremely dangerous. We know that certain diseases, such as malignant cancers, are recessive. That is, the pain or other effects of a tumor that are noticed by a person may disappear for a short time, only to return again. This does not mean that the tumor is gone, but just that the symptoms disappear for short periods of time. If a person with such a disease eats a certain food at a time when the symptoms are disappearing naturally, they may attribute the cure to this food. If the symptoms and pain return, more of the food may be ingested, and if the symptoms go away naturally again, the belief that the food has curative powers is reinforced. When the pain becomes intolerable after a year or so, the person may then go to a physician and find that he has a cancer that could have been cured had he seen the doctor a year ago. In such a case, the use of food as a cure-all might take a life or irreparably harm someone's health.

WEIGHT CONTROL

Weight control is a national obsession, and although it was discussed in some detail in Chapter 1, a few summary remarks might be appropriate.

Weight loss should not be transitory; one's diet should not change every other month. In order to maintain a healthy weight, one should first review all eating habits and establish for life a healthy pattern of eating—a whole new pattern, if necessary. For example, people must stop rewarding themselves with food, associating food with love, and consuming food whenever they feel bad or good. Eating is an enjoyable experience, but it should be placed in its proper perspective when it begins to affect health in the form of overweight and obesity.

One pattern of eating is typical of all the cure-all diets we have discussed in this chapter: the exclusion of certain foods and the promotion of other foods. In other words, if you want to maintain a certain weight, you eat more of a certain food. This means that you must exclude other foods. There are very sound, scientific reasons for excluding certain foods from the diet at certain times. For example, people who have a history of heart disease in their family, suffer from high blood pressure, and have high cholesterol and high triglyceride levels in their blood should lower their intake of cholesterol and therefore those foods that contain it. People who suffer from hypertension should exclude salt or, at most, use moderate amounts of it in their diet. However, in weight loss, as in weight maintenance, a variety of foods provides the greatest safety factor for good health. Knowing something about the composition of foods allows us to choose nutrients dense foods, which have a high ratio of nutrients to calories, rather than the reverse situation.

The final question to be answered about weight is, "How much should I weigh?" Life insurance tables may not be indicative since they represent only those people who buy life insurance. Further, weight in and of itself may not be as important as body composition indicating the percent fat on your body. Nevertheless, people want to have a number or range to aim for and it seems that one of the best measure may be the "body mass index" (BMI). This includes both weight and height and therefore is more meaningful than weight alone. It is calculated by the following equation.

$$\text{BMI} = \frac{W}{H^2}$$

where W = weight in kilograms and H = height in meters.

The desirable range of BMI for men is 20–25 and for women 19–24. If you are above, lose some weight; if you are below, gain some weight.

Since all of you are not familiar with kilograms and meters you can convert as follows:

$$\frac{\text{pounds}}{2.2} = \text{kilograms} = W$$

$$\frac{\text{inches}}{39.37} = \text{meters} = H$$

Let us take an example. A 20-year-old male weighs 150 pounds and is 5 feet 8 inches (68 inches) tall

$$W = \frac{150 \text{ pounds}}{2.2} = 68.2 \text{ kilograms}$$

$$H = \frac{68}{39.37} = 1.72 \text{ meters}$$

$$\text{BMI} = \frac{W}{H^2} = \frac{\text{W}}{\text{H} \times \text{H}} = \frac{68.2}{1.72 \times 1.72} = \frac{68.2}{2.96} = 23.04$$

This individual is within the range for men and, therefore, his weight is okay.

CHOOSING A HEALTHY DIET

This is not a difficult question to answer; we have done it several times thus far in the book. A healthy diet is a varied diet. Checking the composition of one's diet is not really a difficult chore, and everyone should compare their diet with their nutritional needs.

For those of you who wish a simple method of changing your diet, a few general concepts might be helpful. Perhaps the healthiest way to eat is to consume a wide variety of foods but to cut down on your total caloric intake. Therefore, in order to keep your calories down or at least at the same level, you should cut down on fats and fat-rich protein foods but increase your consumption of carbohydrate foods, such as grains, potatoes, fruits, and vegetables. As we noted in Chapter 1, the diet of many Americans is slightly deficient in fiber. The ingestion of more unrefined carbohydrates, such as grains, fruits, and vegetables, will provide the necessary fiber, along with the calories you will need to replace the ones that are lost. Keep in mind also that some Americans have been found to be deficient in iron, vitamin A, vitamin C, and certain B vitamins. You can use nutritional labeling to your advantage by keying in on foods that are high in iron, vitamin A, and B vitamins. In order to obtain your daily requirement of vitamin C, you merely have to drink 4–6 ounces of an accepted breakfast drink that is either fortified with vitamin C or contains it naturally.

The basic four food group plan may also be used to help one to choose an adequate diet. It recommends two servings from the meat group (meat, fish, poultry, eggs, nuts, and beans), two from the milk group, four from the vegetable–fruit group (including one good source of vitamin C per day and one of vitamin A every other day), and four from the bread–cereal group. More recently a fifth group, the fats, sweets and alcohol, was added. This is a reasonably simply system, which builds a foundation for good nutrition, but it could present some problems if foods were chosen unwisely or the system misused.

However, utilizing our new-found knowledge, common sense, and the basic four food group we can apply the following five recommendations to our lives to produce a healthier and happier way of living (Clydesdale 1984).

1. Reduce calories to lose and then maintain a weight consistent with age, sex, body size, and shape.
2. Increase exercise up to a moderate level.
3. Within the confines of the maintenance calorie level, choose a variety of foods, which includes an increased consumption of

grains, fruits, and vegetables and have an appropriate nutrient/calorie ratio to meet the RDA.

4. Stop smoking.
5. Reduce alcohol consumption to no more than 100 calories a day. If this is impossible, quit drinking all alcohol.

HOW GOOD IS MY DIET

Readers who wish to determine more precisely the composition of their diet may do so with the help of the following books, among others.

1. Nutritive Value of American Foods in Common Units, Agricultural Handbook No. 456. Agricultural Research Service, United States Department of Agriculture, Washington DC.
2. Nutritive Value of Foods, Home and Garden Bulletin No. 72, United States Department of Agriculture, Washington, DC.

These books are inexpensive and may be obtained from the extension service of your state university, the Government Printing Office, Washington, DC, or your local or University library.

In order to determine the composition of your diet, the following exercise may be used. It is really quite manageable and the information obtained is well worth your while. For instance, the exercise stops you from fooling yourself about weight control. It also provides you with a "health check" on the adequacy of your diet by enabling you to determine whether you are deficient in any nutrients.

1. Record the following informatin.
   a. Record all food eaten, using Table 5.1 as an example.
   b. Using Table 1.1 determine your Recommended Daily Dietary Allowance (RDA) for each of the nutrients shown (see Chapter 1).
   c. Using one of the two books listed above, determine the nutrients that are provided by the foods recorded in Table 5.1, and enter these results in Table 5.2.
   d. In Table 5.3, record your physical activities during your working hours. Using Table 5.4, calculate the calories expended in your activities, and record the results in Table 5.3.
   e. Calculate your calorie expenditure during sleep by multiplying your body weight in kilograms (your weight in kilograms is equal to your weight in pounds divided by 2.2) by the number of hours of sleep.

Table 5.1. Record of Food Intake for One Day

Name: ______________________________

Food	Measure	Amount in Grams
Breakfast		
Lunch		
Dinner		
Snacks		

Table 5.2. Intake of Selected Nutrients for One Day

Date____________________

Food	Weight or Measure	Food Energy (cal)	Protein (g)	Fat (g)	Carbohydrates (g)	Ca (mg)	Fe (mg)	Vit A (IU)	Thiamin (mg)	Riboflavin (mg)	Niacin (mg)	Ascorbic Acid (mg)
Total Nutrient Intake												

Table 5.3. Estimate of Total Activity for One Day

Time of Day	Form of Activity	Time Spent		Calories Expended
		Hours	Min	
Total calories expended while awake				

Table 5.4. Metabolic Costs of Various Activities, Including Basal Metabolism and Specific Dynamic Action

Activity	Cal/pound per hour[a]
Archery	2.052
Bicycling on level roads	1.997
Bowling	2.653
Calisthenics	1.997
Canoeing, 2.5 mph	1.200
Canoeing, 4.0 mph	2.800
Carpentry	1.534
Chopping wood	2.996
Classwork, lecture	0.667
Cleaning windows	1.652
Conversing	0.735
Cross-country running	4.435
Dancing, fox trot	1.769
petronella	1.853
waltz	2.041
rumba	2.759
eightsome reel	2.721
moderately	1.665
vigorously	2.261
Dressing	1.268
Driving car	1.192
Driving motorcycle	1.445
Driving truck	0.931
Farming, haying, plowing, with horse	2.664
Farming, planting, hoeing, raking	1.867
Farming chores	1.535
Gardening, digging	3.714
Gardening, weeding	2.346
Golfing	2.155
House painting	1.399
Ironing clothes	1.706
Lying, quietly	0.531
Making bed	1.556
Metalworking	1.399
Mopping floors	1.809
Mountain climbing	4.000
Pick-and-shovel work	2.664
Pitching horseshoes	1.409
Playing baseball (except pitcher)	1.867
Playing football (American)	3.205
Playing football (association)	3.559
Playing ping-pong	1.540
Playing pushball	3.053
Playing squash	4.141
Playing tennis	2.759
Repaving roads	1.997
Resting in bed	0.473
Rowing for pleasure	1.997
4.60 mph	3.298
5.18 mph	3.740
5.80 mph	4.536
Running long distance	5.994
Running on grade (treadmill)	
8.70 mph on 2.5% grade	7.216
8.70 mph on 3.8% grade	7.627
Running on level (treadmill)	
7.00 mph	5.564
8.70 mph	6.185
11.60 mph	7.834
Shining shoes	1.189
Shooting pool	0.814
Showering	1.268
Sitting, eating	0.555
normally	0.479
playing cards	0.571
reading	0.479
writing	0.729
Sled pulling (87 lb), 2.27 mph	3.379
Snowshoeing, 2.27 mph	2.272
Sprinting	9.314
Stacking lumber	2.329
Standing, light activity	0.969
normally	0.561
Stonemasonry	2.531
Sweeping floor	1.456
Swimming (pleasure)	3.956
backstroke, 25 yd per min	1.540
backstroke, 30 yd per min	2.117
backstroke, 35 yd per min	2.721
backstroke, 40 yd per min	3.325
breast stroke, 20 yd per min	1.916
breast stroke, 30 yd per min	2.873
breast stroke, 40 yd per min	3.831
crawl, 45 yd per min	3.477
crawl, 55 yd per min	4.234
sidestroke	3.325
Truck and automobile repair	1.665
Volleyball	1.374
Walking on level (treadmill)	
2.27 mph	1.396
3.20 mph	1.877
3.50 mph	1.994
4.47 mph	2.637
Walking downstairs	2.656
Walking upstairs	6.911
Washing and dressing	1.039
Washing and shaving	1.140

[a] Calories used for each pound of body weight doing the activity listed for an hour. To obtain the calories you used multiply this number by your body weight. Modified from Wilson *et al.* (1967).

2. Summarize your results as follows:
   a. Add up the amounts of each of the nutrients listed in Table 5.2.
   b. Compare your actual nutrient intake, obtained from Table 5.2, with the RDA, which you determined in step 1b. This comparison allows you to determine if your daily diet is adequate and, if not, which nutrients you require.
   c. Add up your calorie expenditure while awake, which is recorded in Table 5.3.
   d. Add the total calories you expended while awake (obtained from Table 5.3) to the calories you expended while sleeping (obtained from step 1e). This provides you with a good approximation of your total calorie expenditure. Compare this figure with your calorie intake (obtained from Table 5.2). This step is very important in weight control, since input should equal output for weight maintenance, and input should be less than output for weight loss. Remember, to lose one pound you must expend 3,500 more calories than taken in. If your calculation shows that you expend 1,000 calories more per day than you take in, then you will have a deficit of 7,000 calories per week, which will result in a weight loss of $7000 \div 3500 = 2$ pounds per week.

BIBLIOGRAPHY

CLYDESDALE, F. M. 1979. Nutritional realities—Where does technology fit? *J. Am. Dietet. Assoc.* **74,** 17.

CLYDESDALE, F. M. 1984. Culture, fitness, and health. *Food Technol.* **38**(11), 108–111.

CLYDESDALE, F. M. 1984. View of a viewpoint: Nutritional considerations. *J. Learning Disabil.* **17,** 450.

COMMITTEE ON DIET, NUTRITION AND CANCER. 1982. Assembly of Life Sciences, National Research Council, National Academy of Sciences, Washington, DC.

FALLON, A. E., AND ROZIN, P. 1983. The psychological bases of food rejection by humans. *Ecol. Food Nutr.* **13,** 15.

FEINGOLD, B. 1975. Why Your Child is Hyperactive. Random House, New York.

FORBES, G. B. 1982. Some influences on lean body mass: exercise, androgens, pregnancy, and food. *In,* P. L. White and T. Mondeike (Editors). Diet and Exercise: Synergism in Health Maintenance. American Medical Association, Chicago, Illinois.

HERTZLER, A. A., WENKAM, N., AND STANDALL, B. 1983. Classifying cultural food habits and meanings. *J. Am. Med. Assoc.* **80,** 421.

KOLATA, G. 1982. Food affects human behaviour. *Science* **218**(12), 1209.

WILSON, E. D., FISHER, K. H., AND FUGUA, M. E. 1967. Principles of Nutrition, 2nd Edition. John Wiley & Sons, New York.

ZURER, P. S. 1984. Drugs in sports. *Chem. Eng. News.* **62**(18), 69.

II

Food Supply

Part I of this book deals with food problems as we have come to know them in the United States, where food supply has never been a real problem. This type of security is rare in many countries of the world today, which cannot or will not produce enough food for their population. Prior to World War II, many countries were food exporters; today there are only six: the United States, Canada, France, Australia, New Zealand, and Argentina. Over 70% of the world's food exports originates in the United States and Canada. It would certainly contribute to better political stability in the world if the list of potential food exporters were much larger.

The capacity to increase food production is a very complicated technological, sociological, and political mix. Food production has been outstripping population growth for the past 20 years, but this overall statistic does not tell the whole story. The developed countries have been increasing their food production rapidly, largely by raising their yield per acre. The developing countries have also been raising their food production, on a per-capita basis, largely by bringing new land into cultivation. This is fine if one does not run out of land; land area is finite and in many countries all the land best suited for food production is already being cultivated. As marginal land has to be utilized, the capital costs to bring land under cultivation increase. Eventually, attempts to farm more land become uneconomic, and at that point future increases in food production have to come mainly from increased yields per acre. Can this be readily accomplished? One has to answer yes! But the problems are far from simple. One can attribute American agricultural success to four broad principles, roughly equal in importance: the use of better varities of plants and animals, the use of more and better fertilizer, the application of better pesticide and crop-protection chemicals, and increased mechanization. The developing countries can certainly apply the first three principles, but except for small areas of relatively low population they cannot increase mechanization to the level used in the United States. For example, in parts of India where the arable land is about one acre per person or less, the adoption of American-style mechanization would displace so many farm workers that the resulting unemployment would be a sociological nightmare. The political consequences are unthinkable. Obviously, the developing countries can mechanize to a greater degree, but labor-intensive countries are probably destined to remain labor-intensive for some time. The adoption of the first three principles is imperative, but this involves considerable technological resources, including sophisticated knowledge and capital. Progress on all four fronts is being made, but too slowly, we fear, to cope with population growth.

Countries with a high population density are experiencing a unique type of migration pressure. The usual pattern is for workers to leave the land and migrate to the cities, since the life is usually easier there. In times of economic distress, a reverse pattern usually takes place, since one can at least be sure of getting enough to eat by returning to the land. In some areas, however, the population is already such that the capacity of the land to produce has been reached, and there is a population "push" instead of the usual "pull" to the cities. Eventually, the people who are pushed may literally have nowhere to go. Such a situation may be compounded by a deliberate governmental policy of maintaining low food prices in order to benefit the urban population. This destroys any motivation by rural dwellers to increase food production. Such a scenario precedes more frequent famines and eventual disaster.

The above trend may lead one to think some situations are hopeless, but such is not the case. There are reasons for real optimism. For example, it is estimated that 200,000,000 acres of potentially arable land exist in the Sudan in Africa. If capital can be supplied for irrigation and land development, this territory could become an important source of food. Similarly, vast areas exist in the Amazon valley in South America, a region which has been called the "future bread basket" of the world. Unfortunately, the technology to exploit this land for optimum food production does not exist today, but surely it can be developed.

One author (F. J. F.) traveled over 7000 miles throughout India on a 2-month tour of food production and processing facilities as a guest of the Indian government. It was startling to note the complete dependence of the Indians on the monsoon rains for production of both food and electrical power. Since it was the dry season (February and March, 1976) this connection was even more striking, yet there was a food surplus. The rains in 1975 were excellent, but not every year will be as good. The officials responsible for food supply planning in India recognize the problems and know the solutions. They have a cadre of exceedingly dedicated scientists who are encouraging irrigation projects that will grow three crops a year instead of one, as well as the construction of storage facilities for surplus food. Processing facilities are being developed to reduce the dependence on what is now essentially a fresh food economy. Finally, they are approaching the population question in a realistic manner.

Family planning centers in India exist in every area. They dispense a variety of voluntary family planning advice. Yet some thought that this type of program was not enough. In April, 1976, Maharashtra state passed a compulsory sterilization law. Public outrage soon caused the law to be repealed but the population pressures were still there. Some westerners have expressed moral outrage that such a law should even be suggested. However, regardless of the moral outrage, let no one be under the illusion that population will not be controlled. It will be—the choice is only whether it will be planned or will occur because of famines induced by natural disasters.

China, the most populous country in the world with a population about one billion, has announced a "one-child" family policy. This concept is encouraged through intensive family planning advice, heavy peer pressure, and even a number of legal constraints. Obviously, China believes that control of population growth is important. On the other hand, Malaysia has announced a policy that encourages a higher birth rate. Canada has a "baby bonus" to encourage population growth. Pub-

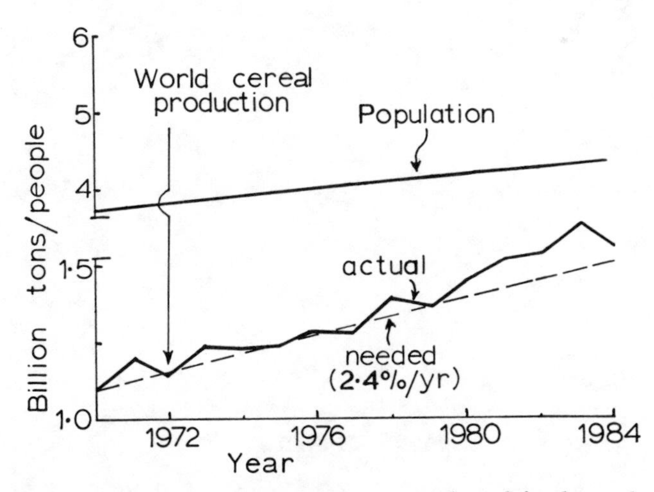

Relationship between population growth and food supply.

lic policy to encourage or discourage population growth obviously differs, depending on the aims and aspirations of each country.

The question is often asked "Is food supply keeping up with population growth?" The answer is an unqualified Yes! The relationship between the two with food supply given as cereal production is shown in the following figure. Population growth is actually slowing down from a worldwide average of 2% per year in 1965 to 1.7% today. A growth rate of 1.7% will double the population in about 41 years. Food supply is increasing at approximately 2.4% per year. This was the figure, suggested by the United Nations, which would allow better expectations for mankind. Yet the overall averages are misleading, since population growth is the greatest in areas least able to cope. There is a great disparity in food supply between areas of the world. Currently, Africa is having great difficulties in feeding its expanding population.

The assurance of an adequate food supply is dependent on a complex interaction of technology, storage, preservation, distribution, capital availability, nutritional considerations, and population policies. Part II of this book attempts to deal only with some of the technological concepts of food supply.

6

Food from Cereals: The Green Revolution

Cereals and root crops provide up to 75% of the total calorie intake in some countries. Eight cereal grains—wheat, rice, corn, barley, oats, rye, sorghum, and millet—provide over 60% of the calories and 50% of the protein consumed by man. There is no question that, directly or indirectly, the cereal grains will continue to supply the major portion of calories in the human diet.

The "Green Revolution" is a general term for a package of technical developments that have doubled or tripled the potential supply of cereal grains. The story reads like a modern miracle, and we need more miracles like this.

WHEAT

Norman Borlaug is known as the father of the Green Revolution. There have been many other contributors, of course, but he received a Nobel Peace Prize in 1970 for his efforts to raise the yield of cereal grains. The story of his life may well inspire other workers who wish to do something significant for mankind.

Borlaug was born on a farm in Iowa in 1914. He received a B.S. in forestry and an M.S. and Ph.D. in plant pathology from the University of Minnesota. In 1941, he joined the DuPont Company in their fungicide testing laboratory. In 1944, the Rockefeller Foundation asked him to go to Mexico to see what could be done about food production there. To understand just what this charge implied, we should look at the food situation in Mexico in 1944. The country was very poor, with three-quarters of the people living on the land. Traditional agriculture was primitive. The land had been worked for over 2,000 years and had received little fertilizer, irrigation, or any other manifestation of what we know as modern agriculture. Farming had a very low social status, and the standard of living for farmers was also very low.

Funded by the Rockefeller Foundation, Dr. Borlaug was free of any political pressures and of any need to show a profit or the usual disadvantages of a short-term assignment. Relationships with the Mexican government and the National Agricultural College at Chapingo, near Mexico City, were cordial but distant. If the program was a disaster, this loose association would not discredit the Mexican government. In this unfettered arrangement, Borlaug proceeded to do what he knew best. He surveyed the wheat situation, and decided that wheat rust was probably the limiting factor in holding down the yields per acre. He collected varieties of wheat from all over Mexico and grew them at five locations within the country. Of the many thousands of varieties tested, most were indeed susceptible to the wheat rust diseases; several, however, were resistant. Painstakingly, Borlaug and his associates made over 30,000 crosses and finally incorporated the resistance to rust into four varities of wheat that made acceptable flour for bread.

The resistance to rust, important as it was, was overshadowed by another aspect of the wheat-breeding program that Borlaug carried on at the same time: the breeding of wheats that were insensitive to photoperiodism. The photoperiod is the biological clock that tells plants when to bloom and when to produce only leaves. For example, chrysanthemums grown outdoors bloom only in late fall when the days are short. The blooming process is initiated by the short length of the days. Similarly, poinsettias, the red-flowered Christmas plant, will flower only if grown during days that are under 9 hours long. If the days are longer, the red "flowers"—actually, they are leaves—will remain green and never turn red. Similarly, a wheat plant with a definite photoperiod will not produce wheat kernels unless the day length is correct. Dr. Borlaug bred the photoperiodism out of his wheat varieties, which meant that a variety could be grown across a latitude of 5,000 miles, rather than 500 miles. It also meant that two or more crops could be grown in the same year.

In 1951 Borlaug had four rust-resistant varieties of wheat, but that same year, race 15B appeared and devastated wheat fields in the United States, Canada, and Mexico. This new race attacked two of Borlaug's rust-resistant varieties. In 1953, a new rust variant appeared and the other two varieties succumbed. The plant breeding research went on, and by 1957, several new varieties had been released and had proved to be rust-resistant. After thirteen years of hard work, the rust problem had been conquered.

What had Borlaug accomplished? Seventy percent of the wheat acreage in Mexico had been switched to the cultivation of the new varieties. Borlaug had almost doubled the wheat yield per acre in Mexico, from 11.5 to 20 bushels. Yet the population growth in Mexico was so great that this increase in food production was not enough.

Borlaug began to study wheat production systems in order to determine what was limiting the yield per acre. It was not disease. It was not water. It was not fertilizer. Increased applications of fertilizer to the existing varieties merely made the plant grow tall and spindly, and the wind and rain soon caused the tall plant to collapse into the mud. The increased applications of fertilizer actually reduced the wheat yield. Borlaug decided that the capacity of the plant itself to use fertilizer was the limiting factor. The answer to this problem came from Japan. The Japanese were masters in the art of dwarfing plants, particularly ornamentals known as "bonsai." Bonsai is the dwarfing of plants by physical means, such as bending limbs and restricting root growth, and has nothing to do with dwarfing by plant breeding methods. But the Japanese appreciated the potential of dwarf plants. The Japanese had ben growing dwarf wheat for 75 years. With these dwarf plants, no matter how much fertilizer—or in earlier times "night soil" (a polite term for human waste)—was used, the plants did not grow tall but merely produced bigger yields of kernels. One variety, Norin 10, which had been released ot Japanese farmers in 1935,was imported by the United States in 1946 and became part of the ancestry of the famous Gaines variety. Borlaug obtained samples of the Norin wheat in 1953 and proceeded to incorporate its dwarfing genes into his previous selections. The results were amazing.

Borlaug developed varieties that could use three times as much fertilizer (from 40 to 120 pounds of nitrogen per acre) and that yielded two to three times as much wheat per acre.

The new varieties changed the entire concept of growing wheat. In order to exploit the capacity of the new wheats to use fertilizer, a number of other inputs were necessary: an adequate supply of water; pesticide treatments to control insects, weeds, and disease; greater amounts of capital and farm credit; and better equipment for harvesting, cleaning, packaging, and transportation. Probably the most important aspect of the development of high-yielding strains was the motivation it provided the growers. A grower at the subsistence level does not have much latitude for experimentation. One mistake can mean disaster. Developments that promise to raise grain yields by 10 or 20 percent are just not worth the risk. However, a development that promised to double or triple yields certainly provided the motivation for change, and it was eagerly accepted. In Mexico, over 90% of the total wheat area had been allotted to the high-yielding varieties by 1967, and the average yield was over 30 bushels per acre.

The new varieties were adopted rapidly by other countries. India and Pakistan imported 1100 lb of the new seeds in 1964 and 600 tons in 1965. During the the disastrous Indian famines of 1965–1966, they had the cure on their own doorstep and they were quick to capitalize on it. In 1966 these two countries imported 36,000 tons of seed grain, and in 1967 another 40,000 tons. By 1969–1970, India and Pakistan had planted over 22,000,000 acres of the new varieties. Both countries had a surplus of wheat in 1971. Never in history had a change of this magnitude been effected so quickly. The story was repeated in Afghanistan, Nepal, Iran, Morocco, Tunisia, Argentina, Chile, Syria, Egypt, Turkey, and many others. In Turkey alone, the acreage devoted to the new wheat varieties rose from 1,500 in 1966–1967 to 1,540,000 in 1969–1970. In India, the development of the high-yielding wheats started with seed from Mexico, but the early Mexican varieties have nearly all been replaced by varieties that are better adapted to local conditions. By 1975, 20% of the wheat acreage in Asia was bveing planted with the new varieties.

RICE

The development of high-yielding varieties of rice was very similar to that of wheat, except that the time span was longer. The story of the development of rice starts in Japan in the late 1800s with experiments to increase the yield of rice by applications of fertilizer. It was found that only the short, erect, stiff-stemmed plants would yield more rice as a result of increased fertilizer. The yield of rice is related directly to the amount of sun the plants are exposed to before they "lodge" or tip over before harvest. When more fertilizer was applied to the conventional varieties of rice, the plants merely grew taller and lodged more quickly, thereby reducing the yield. When Japan turned to occupied Taiwan prior to World War II in order to feed the Japanese homeland, it soon became obvious that the yield of Taiwanese varieties could not be increased by extra fertilizer. The Japanese imported dwarf types from Japan and started to develop varieties that were adapted to conditions in Taiwan. After World War II, when the Chinese regained control of Taiwan, the research continued. By 1956, a variety had been developed that increased yield by about 50%, but few people knew about it.

In the late 1950s, the Rockefeller and Ford Foundations teamed up to share the cost of rice research in Asia. They selected the College of

Agriculture at Los Banos in the Philippines as a base of operations, and there they founded the International Rice Research Institute (IRRI) in 1960. The staff started with three Taiwanese semidwarf varieties and by 1966 had produced IR-8, the most famous rice variety of all time. Conventional varieties produced about 1.5 metric tons of rice per hectare (2.47 acres), whereas IR-8 was capable of producing 6 tons under good conditions. The new "miracle" rice, as it was called, could triple the yield, but it was not universally accepted. The chubby, chalky grains tasted different from the conventional rice, and in spite of the promise of instant plenty, consumers resisted. Another problem was that IR-8 was susceptible to some diseases. After 1966, however, a number of new varieties were developed that are far superior to IR-8. They combine disease resistance with acceptable appearance and taste. The stage was set to revolutionize the production of rice in the rice-eating world.

Despite their potential for greatly increased yields, the new varieties of rice have not ben accepted as widely as the developers had hoped. The miracle rice grows best and yields most under ideal conditions, one of which is controlled year-round irrigation. But two-thirds of the Asian rice land depends on rainfall. Only 25% of it is irrigated, and most of that land does not have dependable water supplies. The spread of the new high-uielding rice varieties will depend on the availability of water, fertilizer, and pest-control chemicals. In 1970, the Asian land estimated to be devoted to the new varieties was about 20,000,000 acres. In 1975, it was estimated that about 10%, 25 million acres, of the Asian rice land was devoted to the new varieties.

Yet in spity of the obvious success of the wheat and rice programs, the Green Revolution has been severely criticized. For example, some say that the rich are getting richer and the poor, poorer. Only the more aggressive, farsighted farmers could take advantage of the potential of the miracle seeds. The cultivation of the new seeds involved irrigation, pesticides, equipment to till the land, farm credit, and so on, so it was only natural that the wealthier and more intelligent farmers would benefit the most. Second, the potential for grain production has made land more valuable, so landowners have atempted to acquire more land and reduce previous owners of marginal land to the status of hired help. Undoubtedly, both claims are true to some extent, but certainly the standard of living has been raised. The tremendous increase in India of tube wells that provide irrigation is evidence that many did share in the increased grain production in that country. Undoubtedly, many farmers, particularly those in the drier areas, are not able to use the new seeds. The third criticism, therefore, is that under these conditions the new seeds may yield even less than the old varieties. This is not true.

Even under por conditions, including a lack of fertilizer, the new varieties usually outperform the old ones. They do not yield up to their potential, but their ratio of grain to stalk is usually better than that of the old varieties, and this does result in more food. A fourth criticism is that now that three times the quantity of grain can be produced on the same land, it is more economical to grow grain rather than pulses (legumes). However, pulses have more and better-quality protein than cereal grains, and if their production goes down, so does the nutritional status of the people.

Critics of the Green Revolution say that the increased use of irrigation water, particularly for rice in the Philippines and wheat in Egypt, has led to an increase in schistosomiasis, sleeping sickness, and other diseases that flourish in water habitats. In Egypt, the Aswan Dam is said to be responsible for a 30% increase in schistosomiasis alone. The parasites responsible for this disease use the snail as an intermediate host, and the irrigation water provided by the Aswan Dam is an excellent growth media for snails. The public health aspects of the increased use of irrigation water have probably not been considered as much as the food production aspects, but the health problems mentioned above can be controlled. Even more damning in the eyes of some critics is that population growth has kept pace with food production increases, due to the new technology.

Another criticism is that there exists the possibility of the emergence of a wheat rust or a rice disease that could attack all varieties that contained, for example, the same dwarfing gene. This could result in a disaster of truly unprecedented scope (see Chapter 13). In 1970, it was estimated that 44,000,000 acres of the new wheats and nearly 20,000,000 acres of the new varieties of rice were planted. The genetic base for both crops was very narrow. Indeed, the entire acreage for both crops was devoted to a very few varieties, all of which had a common origin. The scientists responsible for the introduction of new varieties, particularly after the devastating corn blight in the United States in 1971, have scrambled to broaden the genetic base for both wheat and rice. The supporters of the new varieties have commented that the genetic base of these plants is not as narrow as the critics would have us believe, and to date there have been no major problems due to a widespread outbreak of plant disease. With each passing year, the danger of famine recedes as more varieties are developed, but the danger can never be really eliminated. Another criticism of the Green Revolution, particularly in the Philippines, is that some of the normal protein supply chains have been disrupted. The new varieties of rice under intensive cultivation with close spacing require fungicide sprays to control some of the fungi.

The sprays also kill the crustaceans that grow in the water along with the rice. A wide variety of small crustaceans forms an important part of the protein supply for humans. If this supply is lost, it must be made up in some other way. Any new technology usually requires adjustments on all sides.

The most compelling criticism of the Green Revolution is that it is a technically sophisticated concept introduced by the developed countries into the developing countries. The technology of irrigation, fertilizer, pesticides, and mechanization requires a great deal of energy and other resources and may eventually be self-defeating for developing countries. They may just be unable to afford it. This may be true if the developing countries attempt to realize the full potential of the Green Revolution, but they can surely use part of the new technology. Full exploitation of the Green Revolution obviously depends on the overall development of the country, and must be accompanied by technological and economic proficiency.

The Green Revolution was never intended to be a panacea. At best, it provided the potential to feed many more people. The yields in most parts of the world are still so low that there is ample opportunity for improvement. The new technology has really been applied only to wheat, rice, and corn. Barley, millet, sorghum, and oats have not been subjected to the same intensive research. Cassava, sweet potatoes, and many other root crops should have the same potential for increased yields. It will be the task of the institutes concerned with tropical agriculture to realize this potential.

FUTURE HOPE

The Green Revolution is a striking example of the success of international cooperation and philanthropy. The first truly international research and training institute was the International Rice Research Institute (IRRI). It was funded by the Ford and Rockefeller Foundations in collaboration with the Philippine government. The second such institute was the International Center for Maize and Wheat Improvement (CIMMYT), funded by the Rockefeller and Ford foundations in collaboration with the Mexican government. Both of these international institutes were so successful that a number of others have been founded. Table 6.1 lists 13 such institutes. Later institutes were funded by the Ford, Rockefeller, and W.K. Kellogg Foundations, the U.S. Agency for International Development (AID), the United Nation Development Program, the Inter-American Development Bank, and local governments. Formal lead-

Table 6.1. The International Agicultural Research Institutes

Center	Location	Research	Date
1. IRRI (International Rice Research Institute)	Los Banos, Phillipines	Rice	1959
2. CIMMYT (International Center for the Improvement of Maize and Wheat)	El Batan, Mexico	Wheat, maize, triticale, barley	1963
3. IITA (International Institute for Tropical Agriculture)	Ibadan, Nigeria	Cereals, roots and tubers	1967
4. CIAT (International Center for Tropical Agriculture	Palmira, Columbia	Beef, cassava, legumes, cereals	1968
5. WARDA (West African Rice Development Association)	Monrovia, Liberia	Rice	1971
6. CIP (International Potato Center	Lima, Peru	Potatoes	1972
7. ICRISAT (International Crops Research Institute for the Semi-arid Tropics	Hyderabad, India	Sorghum, millet, peas,	1972
8. IBPGR (International Board for Plant Genetic Resources)	FAO, Rome	Plant genetic material	1973
9. ILRAD (International Laboratory for Research on Animal Diseases)	Nairobi, Kenya	Trypanosomiosis, Theileriosis	1974
10. ILCA (International Livestock Center for Africa)	Addis Ababa, Ethiopia	Animal Production	1974
11. ICARDA (International Center for Agricultural Research in Dry Areas)	Lebanon and Syria	Sheep, cereals, mixed farming	1976
12. IFPRI (International Food Policy Research Institute)	Washington, DC.	Food Policy	1975
13. ISNAR (International Service for National Agricultural Research)	The Hague, Netherlands	Research planning	1979

ership of the Green Revolution has now passed from the Rockefeller and Ford Foundations to an international group of nations and foundations knows as the Consultative Group on International Agricultural Research. The development of these institutes provides a superb mechanism for the transferance and development of knowledge absolutely essential to providing a better food supply. The international institutes have an awesome challenge, and it is hoped that the most recent ones can duplicate the spectacular success of the first two.

BIBLIOGRAPHY

BORLAUG, N. E. 1971. The Green Revolution, Peace and Humanity. Selection No. 35. Population Reference Bureau, Inc., Washington, DC.

PAARLBERG, D. 1970. Norman Borlaug—Hunger Fighter. Bulletin PA 969, Foreign Economic Development Service. U.S. Department of Agriculture, Washington, DC.

SEN, S. 1970. A Richer Harvest—New Horizons for Developing Countries. Orbis Books, Maryknoll Press, New York.

SEN, S. 1975. Reaping the Green Revolution. Orbis Books, Maryknoll Press, New York.

7

Food from Animals

Figure 7.1 depicts a rough classification of the world in three ways: population density, areas of hunger, and the green belts. The hunger area is particularly interesting in that it is centered near the Equator in an area ranging from 10° south latitude to 30° north latitude. Population is also high in, but not confined to, this area. On the other hand, the "green belt" is a band from 30° to 55° latitude in both the Northern and Southern Hemispheres. Unfortunately, most of this belt in the Southern Hemisphere is ocean. Thus, most of the grain production for human use is grown in the green belt of the Northern Hemisphere.

Why are the hunger areas confined to the warmer parts of the earth north and south of the equator? Is it because these areas do not have an appreciable animal industry, possibly due to an inability to control animal diseases? Perhaps the warmer climate makes the control of plant diseases more difficult. Certainly, it is not a lack of photosynthesis, that creates these hunger areas, since the extra sunshine and temperature make the rate of production of organic matter much greater than in the colder climates. Hunger is not exclusive to the warmer climates, since some areas of South America are relatively well fed, but these areas have low population densities. Whatever the actual reason for the hunger, such areas are usually associated with an inefficient animal industry—inefficient because even though 70 developing countries have 60% of the world's livestock and poultry, they produce only 22% of the world's meat, milk, and eggs. Stated in another way, the *developed* countries have 21% of the cattle numbers, but they produce about 50% of the beef and about 50% of the milk (Anon. 1975).

American consumers have developed an insatiable liking for red meat, consuming about 103 pounds per capita in 1971. This appetite for

Fig. 7.1. The world showing hunger areas, green belts, and population density.

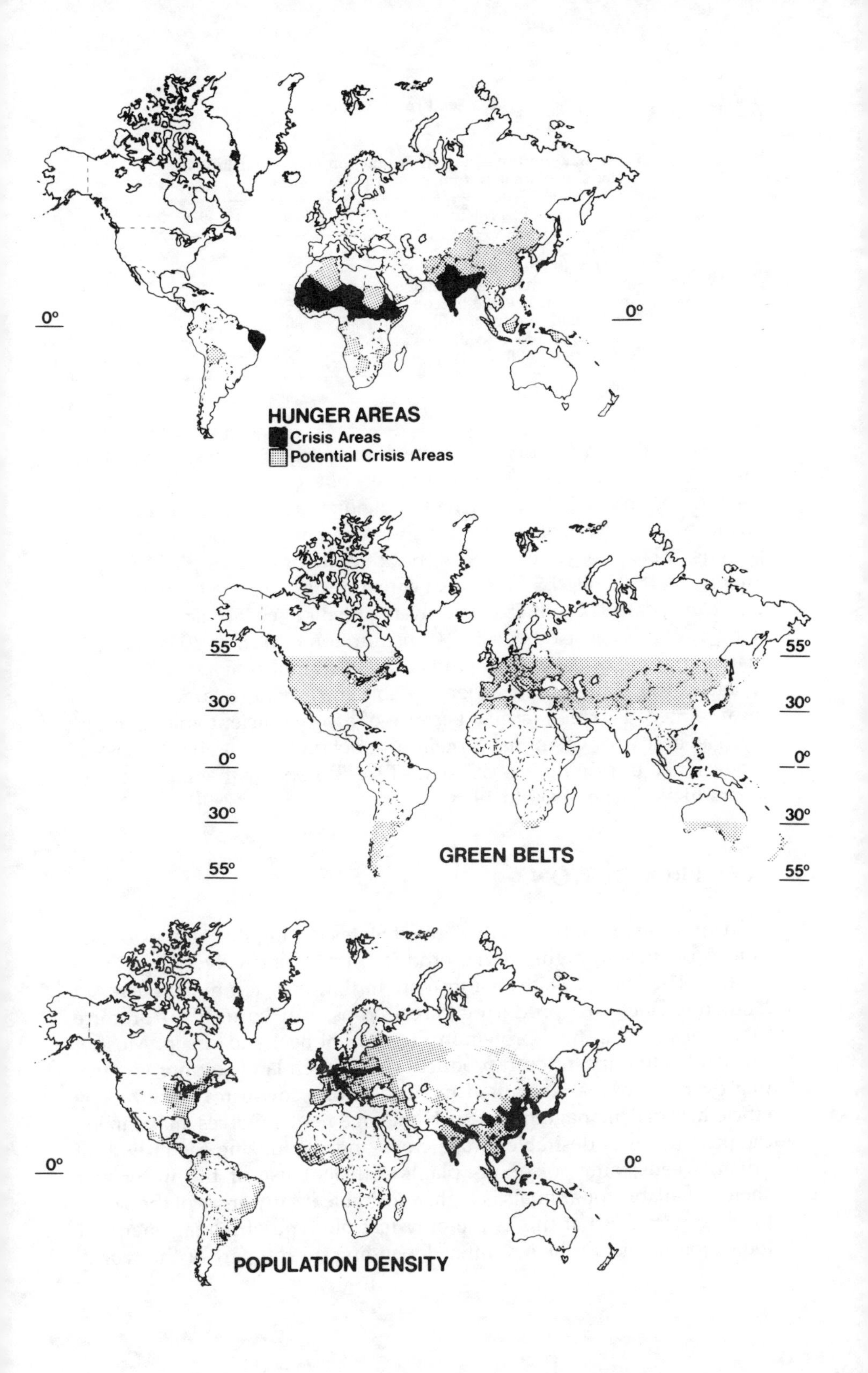

0°
0°
HUNGER AREAS
Crisis Areas
Potential Crisis Areas
55°
30°
0°
30°
55°
55°
30°
0°
30°
55°
GREEN BELTS
0°
0°
POPULATION DENSITY

Table 7.1. Improvement in Production of Animal Protein in the United States from 1930 to 1975

Type of Protein	1930	1975
Annual egg production per hen	121	232
Chicken (lb chicken/100 lb feed)	20	50
Animal egg production per hen	121	232
Turkey (lb turkey/100 lb feed)	15	31
Annual milk production (lb/cow)	4500	13000
Pork (lb per 100 lb feed)	23	35
Beef (lb per 100 lb feed)	8	17

red meat is also on the increase in the developing countries. Apparently, this food preference is related to the level of economic development, at least in countries that do not have religious taboos against red meat. Japan, traditionally considered a fish-eating nation, consumed about 9 pounds of beef per capita in 1974, but the rate of consumption is rising rapidly. In the United States, meat consumption dropped somewhat in the years from 1972 to 1975 primarily as a result of economic constraints, not because of conscious desire. Consumption of beef in 1983 was estimated at approximately 99 pounds carcass weight (see Table 7.5). Fortunately, the United States today is able to provide a large supply of meat to its consumers, largely because of a very efficient animal industry. The efficiency of animal protein production has not always existed; it developed during the last 50 years. Table 7.1 shows the improvement in production from 1930 to 1975.

BEEF PRODUCTION

The production of beef in the United States is a highly complex operation. Any efficient high-volume production of beef involves large areas of land. The United States is fortunate in that over ten million acres of land, unsuitable for producing cereal crops, can be used to produce highly desirable animal protein in the form of beef and sheep. Most of this land is too dry for crop production, but even land so unproductive that over 25 acres are required to support one cow can be utilized by efficient range management. Such management involves maximizing the production of desirable forage crops and minimizing the growth of brush, weeds, and poisonous plants. Efficient use of the water and shelter available for animals is highly desirable. Many areas of the world today are so dry that this is the only feasible type of management for food production. Problems differ depending on the part of the world.

For example, much of Africa could be used much more efficiently if the dreaded trypanosomiasis (sleeping sickness) spread by the tsetse fly could be controlled. In the United States, lands fit only for grazing support about half of all beef cattle and three quarters of all sheep.

Let us examine the production of the typical steak that we like to grill on our backyard barbecue. The process starts on the western plains with the birth of a calf. One cow produces one calf per year. It would be nice if a way could be found to persuade a cow to produce more than one calf per year, but so far, no luck. (The fertility pills show promise, but some problems remain to be solved, particularly the problem of different sexes among the multiple births.) As of now, a cow must be maintained on the range for one year to produce a calf. The calf grows on range forage for six to eight months, at which time it weighs 400 to 500 pounds. It is then moved to pastures that produce more and better grass than the open range, or it is shipped directly to the feeding lots. The feeder lots take animals at 400 to 600 pounds and hold them for three to six months, at which time they weigh 1,000 to 1,200 pounds. The animals are then sent to the slaughterhouses.

The beef produced in the feeder lots is of top quality and is usually assigned one of the top three grades (prime, choice, or good). The main reason for this is that the animal is fed nutritionally balanced rations for optimum growth, such that it reaches slaughtering size as quickly as possible. Assuming comparable sizes, the younger the animal, the more tender the meat. In addition, the feeder lots (100 by 200 feet) restrict the movement of the animals to some extent and thereby use less energy in physical activity. Beef from animals that are confined is usually more tender, and the smaller area makes the feeding process less expensive. The feeder-lot operation is usually very large, in order to obtain the economics of large-scale feed purchases. The large amounts of capital involved in such an operation justify utilization of the latest nutrition information, disease control, computerized inventory control, mechanized handling of materials, and so forth. In short, a feeder lot is a very complex agribusiness.

The calves produced for meat production are obviously of both sexes. Some female calves are selected to replace the breeding stock, and the surplus are sent to the feeder lots. Nearly all the male calves are used for beef production, since in these days of artificial insemination very few bulls are required to reproduce the species. The male calves are always castrated—by removal of the testes—early in life so that they will be easier to handle. A castrated male animal is called a steer, and nearly all of the beef available on the American market is produced from steers or heifers (female calves). Cows and bulls past their productive years are also slaughtered for meat, but such meat is of lower quality and is not normally found on the retail market, except perhaps as small quantities of hamburger, sausage, and other processed meats.

THE DES STORY

The development of the feeder lot has spurred a great deal of research in optimum animal nutrition, disease control, genetic improvement, and other areas. One development has backfired, the use of sex hormones to produce beef more economically.

The use of sex hormones to improve the efficiency of animal protein production is an interesting story. It starts with the unraveling of the chemistry and physiology of the sex hormones during the 1930s and 1940s. When the structures of the female sex hormone estradiol and the male hormone testosterone, and many similar compounds were determined, it was found that a compound called diethylstilbestrol (DES) could be synthesized and made available at a cost much less than that of estradiol. The compound was very similar in physiological effect to the naturally occurring estradiol, but it was not identical in structure. DES was soon made available in large quantities for physiological research and therapeutic applications. Animal physiologists discovered that DES and many similar compounds, including the naturally occurring estradiol and testosterone, had an effect on the rate of growth of animals and the efficiency of food use (see Tables 7.2 and 7.3). Efficiency in this sense is defined as the weight of food consumed, divided by the gain in weight of the animal. The lower the resulting figure, the more efficient the food use. The poultry producers were the first to exploit sex hor-

Table 7.2. A Comparison of Weight Gain and Feed Efficiency of Bulls and Steers With and Without Hormone Treatment

	Steers		Bulls	
	Control	Treated[a]	Control	Treated[a]
Average daily weight gain in pounds	2.25	2.91 (+29%)	6.69	2.73 (+1.5%)
Feed efficiency (pounds of feed/pounds gained)	6.78	6.26 (+8.3%)	6.14	6.21 (+1.1%)

[a] The treatment consisted of 10 mg per day of Synovex H in the feed. Synovex H is a trade name for a mixture of the sex hormones, estradiol and testosterone.
Source: Preston (1972).

Table 7.3. The effectiveness of DES in Increasing Weight Gain and Feed Efficiency of Steers

	Control	Treated (20 mg DES per day)
Average daily weight gain in pounds	2.28	2.86 (+25%)
Feed efficiency (pounds of feed/ pounds gained)	7.24	6.46 (+11%)

Source: Preston (1972).

mones for feed efficiency. They had been producing "capons" by castrating male chickens in order to make their meat more tender. However, since the testes of a chicken are in their body cavity and not easy to get at, less complex means of castration were obviously desirable. Feeding male chickens a mixture of DES and feed was found to produce a "chemically caponized" bird, which came to be known as a "caponette." In 1947 the FDA approved the use of DES pellets implanted behind the head in the neck of chickens, and the process worked—it produced tender chickens.

Several years later the reactions started. Mink producers who were feeding their animals chicken heads complained that their male minks were no longer breeding adequately. A New York newspaper reported that a restaurant worker who was taking chicken necks home for his own use developed female characteristics, including full-sized breasts. Apparently, some of the DES pellets were inserted improperly and were not completely absorbed into the chicken and remained in the neck. The FDA barred the use of DES in 1959. Actually, this incident was not the reason for the ban but the Delaney Clause, which prohibited the addition of carcinogens to food. There was evidence that 6.5 parts per billion (ppb) of DES caused tumors in mice. DES was clearly shown to be present in chicken fat and hence was banned.

DES was approved for use with cattle in 1954, but with the proviso that the additive be withdrawn from the feed 48 hours prior to slaughter. The rationale for allowing DES to be used in apparent violation of the Delaney Clause was that no residues could be detected in the meat. Analytical methods at that time were sensitive to about 50 ppb of DES. Methods were soon developed that were sensitive to 10 ppb, and then 2 ppb. Finally, a radioimmunology method that could detect DES in parts per trillion was created. In the late 1960s, the USDA and other laboratories started to find DES in 0.5 to 1% of their samples of beef liver, in average amounts of 2 ppb. Some samples were reported to be as high as

37 ppb. The DES was found only in the liver, never in the meat. Two ppb is equivalent to 1½ drops in 25,000 gallons. The FDA changed the time of withdrawal for DES from 2 to 7 days, yet minute residues of DES were still detected in livers. One reason may have been the inability of the industry to enforce the withdrawal times.

Public clamor to ban DES developed in the late 1960s, yet the *coup de grâce* was probably due to a completely unrelated event. Three researchers at a Boston hospital reported seven cases of a rare form of vaginal tumor in young women. The only thing the women had in common was that their mothers had participated in an experiment in Chicago from 1950 to 1952. In the 1940s, DES was being prescribed by doctors for women with histories of problems in pregnancy. The Chicago experiment was designed to determine whether DES was of any value in helping women carry a baby to full term. The experiment involved about 1,600 women, of whom half were given 25–100 mg per day of DES during the prenatal period and half were given an inactive preparation (placebo). Neither the women nor the attending doctors knew who received what. This is termed a "double-blind" experiment. The experiment showed that DES was clearly of no value to high-risk-pregnancy mothers and most physicians stopped prescribing it.

This is an example of a well-designed scientific experiment that yielded basic useful information. Yet, this particular experiment was criticized 20 years later because the participants were not given "informed consent," that is, they were not apprised of the possible risks involved. One should consider, however, how many millions of women were saved from the effects of DES because of this experiment.

In 1980, Departments of Public Health, both State and Federal, launched a massive education campaign to alert potential victims. It was difficult to estimate the number of mothers who had been given DES because not all medical records were available; cooperation of the physician involved was sometimes difficult, possibly for fear of lawsuits. Fortunately, the worst fears of tumor formation did not develop. Apparently, the agranulosis condition in the DES daughters was self-healing in many cases. The current estimate of risk is that approximately 1 in 5,000 of the DES daughters will develop tumors in the reproductive tract. There are also indications of problems such as "incompetent" uterus, unrelated to cancer. Also, there are indications that sons of DES mothers may also be affected by problems such as cysts, undescended testes, urinary problems and microphallus (a medical term for a penis less than 1½ inches). However, the evidence for male problems is much less conclusive and the risk is much lower.

The association of DES with potential tumor formation a generation later is most unfortunate. It is unlikely that the 2 ppb found in 1% of beef livers can cause the same problems as the massive doses (up to 1 gram per day) given to the mothers, yet that fact was not appreciated in the press at the time. One might also consider that one type of "morning-after" birth-control drug involves two pills per day, each containing 25 mg of DES, for 5 days—a total of 250 mg. These ten pills have as much DES as can be found in 250,000 pounds of liver. Fortunately, the use of morning-after pills containing DES has lost its popularity.

The FDA banned the use of DES in cattle feed in 1972 and followed, in 1973, with a ban on DES implants. The bans were ruled invalid in 1974 on a legal technicality and reinstated in 1979. The use of DES in animals is illegal today, but a number of preparations involving natural hormones are legal. They are much more expensive than DES. Some are identical with the natural female (estrogen) and male (testosteone) sex hormones, and others are chemically related. Compounds that exhibit activity similar to the sex hormones (estrogens) are widely distributed in the plant kingdom. Grains, soybeans, vegetable oils, and green leafy vegetables are good sources (Anon. 1977).

The use of DES is based on the fact that a steer will reach a marketable weight of 1,200 pounds 35 days earlier, saving about 500 pounds of feed. Other advantages are an increase in the rib-eye area of roasts and steaks, a decrease in the amount of fat developed in the animal, and a decrease in manure production. The savings to the consumer are estimated to be from 3½ to 15¢/lb for steak at the retail level. But, several consumer advocates have asked, "In view of the unknowns, do we really need this saving?"

If we assume that all beef liver on the market contained 2 ppb of DES (a gross overestimate), the average per capita consumption of liver would contain a daily dose of 0.01 nanograms of DES. Compare this to the naturally occurring estrogenic activity in humans: A 50 lb boy would have 40 to 1400 times as much estrogenic activity; a 120 lb woman would have 2500 to 25,000 times as much. (See Chapter 13).

As an alternative to using DES to increase growth rate, why not raise bulls instead of steers? Bulls do not show any increase in growth rate from DES, since they already grow quickly and are efficient in feed use. An obvious reason for not doing so is that bulls are hard to handle in mixed lots. However, if they are raised in one lot, they are less prone to fighting and can be handled with reasonable care. Another reason is that the American consumers are not accustomed to eating bull meat. Probably the only such meat they have eaten is that of breeding bulls that are slaughtered at an advanced age, a procedure that is guaranteed to produce tough meat. A third reason for not raising bulls is that the meat-grading system has traditionally placed bull meat in a separate category. A new class of meat, "bullock beef," became available in 1974 covers bulls under two years of age, which are graded in a manner similar to that of steers. Bullock beef is leaner, more variable, and probably tougher than steer beef but otherwise quite desirable.

The production of top-grade beef in feeder lots has been criticized because it requires larger amounts of grain. It is also expensive owing to the recent rise in the price of feed grains. The producers have responded by reducing the time that the animals spend in the feeding lots and increasing the time they spend on the range. In recognition of this decrease in feeding-lot time, the "choice" grade has been changed to include beef with less fat than before. These developments will reduce the cost of beef, produce meat with slightly less fat—which may be good for the American consumer—and make the beef slightly tougher. There is also a movement to market more cattle that are range-fed according to the European methods of beef production. Beef from range-fed cattle is even leaner and tougher, primarily because these cattle take a much longer time to reach marketable size. To grow a beef calf from weaning weight of 450 pounds to a slaughter weight of 1050 pounds requires 24 months or more when fed on roughages but only about 8 months when fed on highly digestible feeds such as grain products.

The beef industry, and the American consumer as well, have been criticized on the grounds that beef production requires grain that could be fed to humans. This is partially true but hardly realistic. The grains that are consumed by humans are primarily wheat and rice, very little of which is fed to cattle. Beef cattle are fed primarily corn. Corn is produced on land that could be used to produce food for humans, but there is no assurance that if Americans did reduce their meat consumption, the released corn would in fact be used for human food. A better case could be made for poultry and pigs, since they do compete for food that humans can consume. On the other hand, cattle on the range do not compete with humans. It is often heard that 1 pound of meat represents 10 pounds of grain, implying that this is all it takes to produce a pound of meat. This simplistic approach is not realistic. A truer picture would be that a 1-pound steak represents 300 pounds of grass, 18 pounds of grain, and 3 pounds of protein supplement. This is based on the assumption that a 1,000-pound steer would consume about 10,000 pounds of grass, 2,500 pounds of grain, and 400 pounds of protein supplement, and would produce about 130 pounds of steak. Another 30,000 pounds of grass, or other roughage, would be required to maintain the mother cow for one year. Obviously, a steer produces more meat cuts than just steak. A 1,000-pound steer produces in all, about 420 pounds of retail meat cuts. Thus, the above figures would be reduced if our calculation were based on a "yield" of 420 pounds of retail meat rather than 130 pounds of steak. That is, one pound of retail meat (out of a "yield" of 420 pounds) would represent less grass, grain, and protein supplement than one pound of steak (out of a "yield" of 130 pounds).

The above calculation depends on the time the animal remains on the range or, inversely, the time the animal spends in the feeder lot. Beef animals grow more slowly on the range because of the way they are nourished. The first stomach, or rumen, of a steer or cow is really a big fermentation vat. The microorganisms in the rumen break down the carbohydrates, fats, and proteins ingested by the animal and use the products for their own growth. The steer then digests the microorganisms as its own source of nutrients. This is why it is possible to feed urea to animals. The microorganisms use this nitrogen to synthesize their own protein, which is eventually digested by the animal. Unfortunately, the amount of urea that can be given the animal is limited, because ammonia is readily produced in the rumen by the action of enzymes on the urea, and this is hard on both the animal and the microorganisms. Simple chemical compounds that release nitrogen more slowly than urea does, such as biuret, show promise of becoming inexpensive sources of nitrogen for the synthesis of protein by microorganisms in the rumen.

There is another fascinating aspect to the operation of the rumen of an animal. One major function of the action of microorganisms in the rumen on carbohydrates such as starch and cellulose is to produce large quantities of simple organic acids, such as acetic, propionic, and butyric acids. The animal utilizes these acids as a major source of energy. The FDA has approved a new additive (monensin sodium) that increases the production of propionic acid in the rumen at the expense of acetic and butyric acids. This focus on propionic acid makes sense, because the formation of the other two acids in the rumen is inefficient: energy is lost in the form of carbon dioxide and methane, which the animal excretes by belching. The addition of 30 grams of monensin per ton of feed allows the animal to grow at the same rate with 10% less feed, thus saving up to 300 pounds of grain per animal. Just how the additive works is not known at the present time.

Despite the considerable progress already made, there is still room for improvement in the efficiency animal production. The ability of the ruminant animals (cattle, goats, sheep, and others) to convert inedible (for humans) organic matter into highly desirable food will ensure the place of these animals in the food chain.

MILK

The cattle industry has one way to produce protein from grasses and grains—the production of milk. The dairy cow is the most efficient in converting energy to protein, followed by the chicken and hen in the United States, then the pig, and finally beef cattle. The efficiency of production is not reflected in consumption patterns. In 1975, about 42% of the nation's protein consumption was provided by meat, 22% by dairy products, and 6% by eggs (Anon. 1975). The American consumer just likes meat.

Milk has always been important to mankind. Apart from its acceptance as a superb food, it has been held responsible for increasing longevity. Indeed, the ten countries with the highest life expectancy, ranging from Sweden (74.2 years), the Netherlands, Norway, Denmark, Canada, France, the United Kingdom, Switzerland, and New Zealand, to Australia (71.0 years), all consume large quantities of dairy products. Whether this is the main reason for the longevity is debatable, but the fact remains that dairy products are a nutritious food.

A modern dairy farm is a complex operation, streamlined to produce milk as efficiently and economically as possible. This involves the selection over hundreds of years for animals that will produce large quantities of milk. The selection has proceeded to the point where a dairy cow will eat four times its food requirement for maintenance of its body and turn the excess into milk. Obviously, the efficiency of the conversion depends on the feed ration for the cow (see Table 7.4). Much higher yields of milk can be obtained, if the rations contain concentrated sources of calories and proteins, namely grains and oilseed protein meal, in addition to forage. The traditional index of efficiency in the

Table 7.4. Production of Milk by Dairy Cows

Feed	Yield of Milk (lb/year)
Forage (no feed concentrates)	8,000
Forage + 25% feed concentrates	12,000
Forage + 50% feed concentrates	20,000
Forage + 60% feed concentrates	30,000

dairy industry has been the total yield of milk and the percentage of butterfat. Unfortunately, this concept is obsolete today, since it is much cheaper to produce fats and oils for human use directly from plants rather than use a dairy cow. Butter has only one advantage over margarine, butter has a better flavor as a spread and in baked goods. Having conceded that, all the other advantages are on the side of margarine. This situation has convinced some dairy scientists that the emphasis in dairy cattle breeding should be on protein production, not fat. This obvious conclusion has been very slow in adoption, and butterfat production still remains the usual basis of payment to producers.

The production of dairy products has yet another advantage in that the operation is associated with the production of beef. About 30% of the beef produced in the United States comes from dairy animals. A dairy cow produces one calf per year, which, if not needed for replacement of the dairy herd, can be sold as a feeder calf for beef production. Feeder calves of both sexes from dairy-type animals are just as suitable for beef production as beef-type animals. Dairy-type animals are also marketed as veal. If they are slaughtered when only a few days old, they are called deacon calves or bob veal. In the United States, calves to be marketed as veal are usually fed whole milk or reconstituted milk for 6 to 8 weeks, at which time they weigh approximately 200–250 pounds. In Europe, veal calves are usually marketed at 500–600 pounds. Consumption of veal in the United States is very low, approximately 4.1 pounds per person in 1975.

The efficiency of feed conversion plus the almost universal acceptance of dairy products will ensure the place of the dairy cow in the future food supply.

PORK

Meat from pigs has been an important food for man for thousands of years. Except for people whose religious beliefs proscribe its use (Moslems and Jews), pork is an important food throughout the world. The modern pig is believed to be descended from the European wild boar and the East Indian pig in the area now known as Iraq and dates back to about 6750 BC. Domesticated pigs were found in China about 4900 BC and introduced into the United States on the second voyage of Columbus in 1493. Apparently they thrived, and salt pork became a mainstay of the early colonial diet.

The raising of pigs in the United States has been described essentially as a means of transforming corn into meat. Corn is the primary carbohydrate and protein feed for pigs, but to this diet must be added high protein supplements, minerals, vitamins, and medicinal preparations. Actually, pigs will eat almost anything, but for optimum growth they require a balanced diet. In some countries, the pig is a true scavenger and is raised as such, but obviously, without the benefit of good nutrition they grow much more slowly. Pigs are monogastric—that is, they have one stomach as compared to cattle, sheep, and goats, which have four stomachs. They cannot use plant roughage containing cellulose as efficiently as the animals with multiple stomachs; thus pigs compete with man for food. This is literally true, but they also consume materials such as waste products of the fish- and meat-processing industries which humans may not consider to be very appealing. Also, a small proportion of pigs in the United States are fed cooked garbage. Garbage to be fed to hogs must, by law, be cooked to prevent the spread of a virus disease (exanthema) and a parasite (trichinosis).

The great advantage of pigs is their ability to produce large numbers of offspring and their rapid growth. A good sow can produce two litters of ten piglets each in one year. With good rations, the piglets grow to marketable weight (220 pounds) in 5–6 months. If overfed—and pigs apparently are quite willing to overeat—they will grow quite fat. In earlier times, lard was a valuable commodity and fat pigs were welcome. However, in the 1930s, when vegetable shortenings became popular, the demand for lard decreased. The animal breeders, recognizing that the American consumer wanted leaner pork, converted the pigs from lard to meaty types. In 1955, packers cut 35 pounds of lard from a 220 pound pig, whereas in 1975, it was only 20 pounds. Research to produce pigs with an even thinner layer of fat around the meat is continuing. There is a limit, however, because it is the layer of fat surrounding the meat that is being reduced, not the fat within the meat (the marbling effect). It is not desirable to reduce the fat within the meat because this would make the pork less juicy and flavorful. Another factor that contributes to the flavor of pork is due to a steroid compound which is a normal metabolite of the male hormones. The meat of male pigs or boars which are kept for breeding purposes has an unpleasant sex odor described by some as "perspiration" odor. Male pigs to be used as human food are castrated at an early age to prevent their meat from acquiring this off-flavor. Women apparently are more sensitive to this odor than men. The odor may make the pork less attractive, but it does not affect the wholesomeness.

The production of pigs in the United States is becoming a specialized agribusiness operation. Some operators specialize in producing young feeder pigs from brood sows. Others specialize in growing the feeder pigs to marketable size. Others specialize in slaughter and merchandizing. The degree of specialization has not progressed as far as has the production of beef, but the trends are similar. With more progress in disease control and waste removal, pork production could become even more highly specialized.

In 1983 Americans consumed about 61 pounds of pork per capita, as compared with 80 pounds of beef. Even though consumer surveys have shown that pork is not considered to be a particularly desirable form of meat, it is economical and will continue to play an important part in the American diet.

POULTRY

The production of highly desirable protein in the form of poultry and eggs is the food production success story of the century. America has been the leader in this development, and other countries are now starting to use our methods. High-quality poultry meat is available at low cost in the United States. Foreign visitors are amazed at the relatively low cost of what is considered a luxury in some countries.

Poultry production involves mainly two types of birds, broilers and roasters. The difference is one of age and, hence, size. Broilers weigh up to 4 pounds, and roasters average 6 pounds. Broilers and roasters are of both sexes. Some male birds, called capons, are castrated and raised to larger sizes, sometimes up to 12 pounds. The castration does not improve the rate of growth but does make meat of much superior quality.

A typical broiler operation is a highly sophisticated vertically integrated agribusiness. A company usually owns a feed supply business, a chick hatchery, and a slaughtering and marketing facility. The integrator provides chicks, feed, medicinals, and some specialized services to the producer, who usually owns the growing facilities and provides the labor. The integrator usually owns the birds at all times and pays the producer for his services. Sometimes, however, the producer owns the birds. In a typical operation, the integrator supplies a given number of day-old chicks to the producer, who feeds them for 50–60 days and ships them back to the integrator at an agreed-upon weight, usually up to 4 pounds. The premises are then completely free of birds for 7–14 days. Then another cycle is started. The all in–all out system is desirable as a means of reducing the spread of disease.

The operation is very efficient in terms of labor, since one person can handle up to 100,000 birds at a time, using automatic feeding, watering, lighting, ventilation, temperature control, and so forth. Some of the operations are very large in order to take advantage of the economics of scale. This makes it possible and indeed essential to utilize the most recent knowledge of poultry nutrition, disease prevention, rates of growth, cost of alternate sources of feed, and so on. The application of modern science has made the chicken a very efficient feed converter. A 4-pound broiler can be produced from 8 pounds of feed. This efficiency decreases as the bird ages; for example, about 19 pounds of feed are needed to produce a 7-pound roaster. The production of roasters and broilers is similar, except for the length of feeding. In addition to broilers and roasters there are roosters, which are surplus male breeding stock, and fowls, which are hens over 1 year of age, and too old for efficient egg production.

Most broilers and roasters are grown in pens on the floor, each of which contains about 2,000 birds. Another system is being developed in which the birds are grown in cages that become the actual containers shipped to the slaughtering plant. The catch-and-transfer operation sometimes bruises the birds, a problem that can be eliminated by the new method, which appears to be the ultimate in automation.

EGGS

Eggs are one commodity accepted by almost everyone throughout the world; consequently they are important as a source of food. The United States produced 68.7 billion eggs in 1969, which works out to about 340 eggs per person per year. Egg production dropped to about 64 billion in 1975 possibly because of the adverse publicity due to cholesterol content. The 1984 production is about 70 billion. The USSR is the next largest producer with 58 billion in 1975. Mainland China is thought to have high egg production, but no reliable data are available.

The production of eggs prior to World War II was largely a farm-flock business, with the majority of flocks having less than 400 laying hens. Today, production is a highly specialized business with many flocks having 30,000 or more hens and some up to 1,000,000. One of the main reasons for the increase is the development of mechanical equipment to aid the caretaker in feeding, watering, ventilation, egg handling, and litter management. One worker can easily care for a flock of up to 30,000 birds.

Modern egg production is divided into a series of specializations. A small number of breeders supply fertile eggs to the hatchery operators. These in turn provide day-old chicks to the firms producing the pullets that are sold to the laying flock managers. Similar specialists produce the rations, which are specially compounded for each phase of the hen's life. Other specialists produce the equipment and housing. Still others are marketers of shell eggs, which go to the consumer as such or in various formulated products. Modern research in disease control, sanitary handling, and poultry nutrition has played a very important part in the success of modern egg production systems.

There are some curious beliefs today about the quality of eggs. One has to do with egg color. In New England, brown eggs are preferred, whereas in the rest of America, white eggs command a premium price. There is no detectable difference in taste or nutritive value between eggs with white or brown shells. Some food faddists would have us believe that fertilized eggs are more nutritious. There is no measurable nutritional difference between fertilized and non-fertilized eggs. Others say that "organic" eggs are more nutritious, but there is no scientific evidence to back up this claim. Regardless of various beliefs, eggs are a well-accepted food and provide high quality animal protein at a low price.

SHEEP

Sheep and lambs have been associated with the progress of mankind at least as far back as 4000 BC. They are popular in many countries, since apparently there are no religious taboos against them. For example, in India, where Hindus do not eat beef and Moslems do not eat pork, sheep and chickens are popular with the non-vegetarian populace.

Sheep, marketed as mutton for mature animals and lamb for younger animals, have been popular in the United States since their introduction by Columbus in 1493. In colonial times, sheep were highly prized as a source of both wool and meat, and they were able to utilize forage in arid areas that were not favorable for cattle. However, their importance is declining. The introduction of synthetic fibers and the importation of wool from other countries with very large herds of sheep (such as Australia and New Zealand) has led to a decreased demand for American wool. American consumption of lamb and mutton (mostly lamb) in 1945 was about 7 pounds per year per person. In 1975, it was about 2 pounds. Therefore, lamb cannot be considered an important part of the diet. Its production in the United States may decline even further, since imports of both lamb and wool will probably increase.

GOATS

Goats were probably the first animal domesticated by man; they date back to the New Stone Age. According to the Old Testament, they were very helpful to man, furnishing milk and meat for food, fiber for clothing, and skin for bottles. Goats and sheep are very similar in that they can forage in arid or mountainous areas unsuitable for cattle.

The distribution of goats in America is interesting. Texas has over 90% of the goat population, and over 95% of the Texas goats are of the Angora breed. Angora goats are grown primarily for their fleece, which is known as mohair. Goat meat tastes much like mutton or lamb and is equally nutritious. However, since they are grown primarily for mohair, goats are not usually fattened as well as sheep; hence the yield of meat is lower. Most goat meat is sold under the trade name of Chevon, but consumption is so low that most consumers do not recognize the name.

Goats are also maintained as a source of milk. A good doe will supply 2 quarts of milk per day for 10 months. Some produce as much as 4 or 5 quarts. Actually, more people in the world consume dairy goat products than cow's milk products simply because goats are more widespread. The consumption of goat milk products in the United States is increasing because goats are more adaptable to small towns and suburbs or low income areas where there may not be enough feed for a cow. Goats are not likely to replace cows in the United States because the labor input per quart of milk is twice as much for goats as for cows.

Consumption of goat's milk in the United States is still very low even though it is a good product. Goat's milk has a higher mineral content and smaller fat globules than cow's milk and it forms a fine soft curd, thereby making it more digestible. Goats seem to be getting more public attention as a source of milk for those interested in the "back-to-nature" movement, but their overall importance in the food supply at the present time is minimal.

CALCULATION OF MEAT CONSUMPTION

The figures for meat consumption are misleading because they may be quoted in terms of carcass weight, retail weight, or as the amount actually consumed. For example, in 1981, the amount of beef available per person was 97 pounds per year. This is carcass weight calculated by the total production, including imports, and is usually called "disappearance" data. When the bones and some fat are removed, the figure is reduced to 68 pounds as retail weight. When the retail trim weight,

Table 7.5. Daily per capita Consumption of Red Meat in Ounces for the Year 1983

Consumer Segments	Fresh Beef: Ground Beef	Fresh Beef: Other Beef	Fresh Pork	Veal	Lamb	Processed Meat	Total
Occasional Users	0.07	0.16	0.06				0.29
Moderate/Light Users	0.44	0.99	0.36			0.97	2.76
Heavy Users	1.22	2.99	1.11			3.36	8.68
Total Users	0.68	1.47	0.41			1.65	4.21
Total Population	0.65	1.42	0.34	0.04	0.03	1.50	3.98

Source: Breidenstein (1984).

cooking losses, and plate waste are subtracted, the amount is reduced to 33 pounds per person per year, or about 1.44 ounces (40 grams) per day of beef, not including processed meat. But again, averages may be misleading. A better way to present this type of data is shown in Table 7.5, which indicates consumption by occasional, moderate/light, and heavy users for 1983.

BIBLIOGRAPHY

ANON. 1973. Alternative Sources of Protein for Animal Production. National Academy of Sciences, Washington, D.C.

ANON. 1975. The Role of Animals in the World Food Situation. The Rockefeller Foundation. New York.

ANON. 1977. Hormonally Active Substances in Foods: A Safety Evaluation. Council for Agricultural Science and Technology Report No. 66.

ANON. 1980. Foods From Animals: Quantity, Quality, and Safety. Council for Agricultural Science and Technology Report No. 82.

BRADENSTEIN, B. C. 1984. Contribution of Red Meat to the U.S. Diet. National Livestock and Meat Board, Chicago, Illinois.

CAMPBELL, J. R., and R. T. MARSHALL 1975. The Science of Providing Milk for Man. McGraw-Hill Book Co., New York.

EMSLIE, W. P. 1970. Fifty Years of Livestock Research. Western Publishing Co., Hannibal, Missouri.

ENSMINGER, M. E. 1969. Animal Science. Interstate Printers & Publishers Inc., Danville, Illinois.

HAYES, J., (editor). 1975. That We May Eat. Yearbook of Agriculture, U.S. Department of Agriculture, Washington, DC.

POND, W. G., MERKEL, R. A., McGILLIARD, L. D., and RHODES, V. J. 1980. Animal Agriculture: Research to Meet Human Needs in the 21st Century. Westview Press, Boulder Colorado.

PRESTON, R. L. 1972. Why hormones for beef cattle? Ohio Report, Jan-Feb, pp. 8–11.

SCHNEIDAU, R. E., and DUEWER, L. A. (editors). 1972. "Symposium on Vertical Coordination in the Pork Industry." AVI Publishing Co., Westport, Connecticut.

STADELMAN, W. J., and COTTERILL, O. J. 1977. Egg Science and Technology, 2nd Edition. AVI Publishing Co., Westport, Connecticut.

Food from the Sea

UNLIMITED POTENTIAL?

Man has always looked to the sea as a source of food, and it has served him well. The promise of almost unlimited protein from fish has led many countries to invest heavily in ships to catch and process fish for human consumption. Traditionally, the Scandinavian and other European countries have had a long history of commercial fishing. Lately, the Soviet Union, Japan, East Germany, Poland, Spain, Iceland, Korea, and other countries have joined the fishing fleets in almost every fishing ground in the world. Peru has the distinction of landing the greatest annual tonnage of fish, but nearly all of it goes into fish meal for animal consumption.

In the 1950s and 1960s, the prospects for harvesting fish seemed almost unlimited. This led to many popular articles and books proclaiming that the sea alone could support five or even ten times the population of the world at that time. In theory this may be true, but in actuality we do not have the technology to harvest that many living creatures from the sea in a form usable by man. There may have been grounds for optimism in the increasing fish landings of the 1950s, since at that time, the world landings of table-grade fish, comprising some thirty species, were approximately 21 million tons.

In 1970, the fish catch was 70 million tons and had been increasing at 8% per year. Then, however, the results of overfishing became evident and the catch levelled off. The 1984 catch is estimated to be 68 million tons from the ocean and 7 million tons from fresh water. The total catch is increasing very slowly, and there is little hope for much increase without stringent conservation measures.

The 1974 conference on sea law in Caracas, Venezuela did not show much cause for optimism. The Caracas Conference was one of the biggest international meetings of its kind ever held. Five thousand delegates from 150 countries labored 10 weeks to come to grips with the problem of anarchy on the high seas. Over 250,000 pages of documents were generated per day in five "working" languages. The only firm decision resulting from this massive effort was to hold more conferences. The fish conservation area was only one of many important issues, but it was evident that the accepted theory was that the fish are there for all to take. It is becoming obvious that this attitude will have to change, and soon. The conferences in 1980 and 1982 on sea law made some progress toward conservation by clarifying the 12 mile coastal zone and the 200 mile economic zone. Now each nation can control the fish catch within 200 miles of its coastline. This was one of the minor topics, however; most of the discussion involved the question of sea-bed mining.

The development of vast fishing fleets has had profound political impact. For example, the Soviet Union has built very large fleets of vessels to catch and process fish in all parts of the globe. She probably has over 16,000 fishing vessels. Apparently, the Soviet government decided soon after World War II to divert tremendous quantities of capital to the building of ships that would supply the Soviet people with protein from fish. The United States, on the other hand, put its faith in the development of animal husbandry that would supply protein from red meat.

The Soviets have not been particularly successful in developing abeef industry, but they are trying now. In 1962, a short grain crop occurred in Russia and not enough feed grains were available to feed the beef herds. The Soviets reacted by slaughtering a portion of their animal population. In 1973, a combination of weather factors resulted in another short crop. This time, the Soviets did not reduce their number of animals but elected instead to purchase 30 million tons of grain from the United States. The resulting decrease in American supplies in a hungry world was alarming. It resulted in worldwide publicity for grain sales and price rises, and a realization that the margin of food supply in the world was dangerously thin. It was suggested that the bleak forecast of the Paddock brothers in the book *Famine 1975* might indeed have come true.[1]

In January 1976, it became apparent that the Soviet feed grain harvest was a disaster once again. The expected crop of 215 million bushels was only 135 million bushels. There was no way to make up for the 80-million-bushel shortfall. The Soviets contracted to purchase about 30

[1]The Paddock brothers wrote a book in 1966 which created quite a stir. It was one of the most pessimistic of the "gloomers and doomers" and predicted widespread famine in 1975. W. Paddock and P. Paddock, *Famine 1975*, Little Brown & Co., Boston, 1967.

million bushels, mainly from the United States, but they did not have the transportation system in the interior of Russia capable of handling imports over 30 million bushels. The Soviet animal population was reduced because of the lack of feed. The increased slaughterings produced plenty of meat during the winter, but shortages developed in the spring and summer of 1976.

The Soviets apparently are reconciled to being permanent importers of food. They can pay for it by exporting oil and gold. It may well be a good decision to utilize their capital to generate foreign exchange and buy food from the efficient producers.

THEORETICAL PRODUCTION

To one who has sailed the Atlantic or the Pacific Ocean, it may seem incredible that one could attempt to calculate theoretically the production of fish in all the oceans of the world. Yet numerous attempts have been made, albeit with a high degree of uncertainty. One of the more credible concepts was developed by J. H. Ryther.

The forecasting of ocean productivity started with the development of a method that used radioactive carbon to measure *in situ* the rate of photosynthesis in marine algae. Surely, in the calculation of fish production the most basic starting point is the uptake of carbon in the open sea. This technique was used by the research vessel *Galatea* from 1950 to 1952 in all the oceans of the world. The *Galatea* made only 194 measurements, one for every 2 million square kilometers, but more recent observations, including 7,000 supplied by the Soviets, have not changed the picture appreciably. The data from this research permit the following conclusions:

1. Carbon fixation in the open sea averages about 50 grams of carbon per square meter per year. The open sea accounts for 90% of the surface of the world's oceans.

Fig. 8.1 Nine trophic levels in the ocean.

Table 8.1. Estimate of Potential Yields per Year at Various Trophic Levels[a]

	Ecological Efficiency Factor					
	10%		15%		20%	
Trophic Level	Carbon	Total Weight	Carbon	Total Weight	Carbon	Total Weight
0 Phytoplankton	19000		19000		19000	
1 Herbivores	1900	19000	2800	28000	3800	38000
2 First Carnivores	190	1900	420	4200	760	7600
3 Second Carnivores	19	190	64	640	152	1520
4 Third Carnivores	1.9	19	9.6	96	30.4	304

Source: Ryther (1969).
[a] Values given in millions of metric tons.

2. The shallow coastal waters (less than 180 meters in depth) average about 100 grams of carbon per square meter per year. They constitute about 9.9% of the ocean surface.
3. A few areas of the world are very productive, averaging 300 grams of carbon per square meter per year. They constitute about 0.1% of the ocean surface.

The areas in the last-mentioned category are waters off the west coast of lands that experience prevailing offshore winds—Peru, California, Africa, Somalia, and the Arabian coast. Surface waters in these areas are blown offshore, and the deeper, colder waters that replace them generally have a high content of dissolved minerals, which, together with sunlight, allow the marine algae to grow. Waters with extensive upwelling also exist near Antarctica, but they are not well charted.

Table 8.1 and Fig. 8.1 depict the food chain in the ocean. Phytoplankton are the primary organisms in the chain. They average about

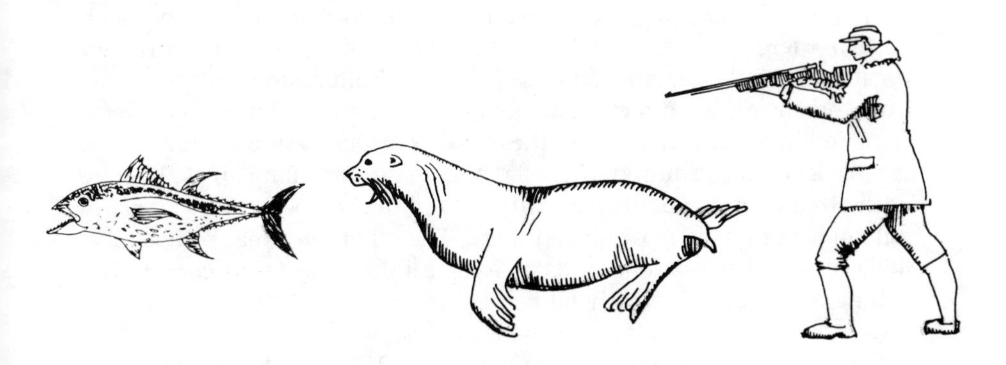

0.0001 inches in size and use sunlight and mineral salts to fix carbon dioxide by photosynthesis. The next stage in the chain, the herbivores, averaging 0.001 inches in size, feed on the smaller plankton. The next stage, the first carnivores, averaging 0.04 inches, feed on the herbivores. The second-stage carnivores, averaging less than one inch in size, feed on the smaller carnivores, and so it goes until a stage usable by man is reached. The concept of one size of organism feeding on another is called a "trophic" level. There is obviously an efficiency factor here in terms of growth. This can be visualized by, say, a fish that has reached full growth eating 1 pound of smaller fish. The full-grown fish does not increase in weight at all, so when it eats fish the efficiency is zero. However, a young growing fish weighing 1 pound may eat 1 pound of fish and gain one-fifth of a pound in weight. This trophic level would have an efficiency of 20%. In fish populations, which, of course, contain both very young and mature individuals, the efficiency ranges from 0 to 20% and perhaps even higher. The 10, 15, and 20% levels are accepted averages. At the 10% efficiency level, for instance, whenever one fish eats a smaller fish—that is, a fish at a lower trophic level—90% of the weight of the smaller fish is lost, returning to the sea in the form of nutrients that go through the carbon-fixation cycle again. Or, this 90% may merely sink to the bottom if the area is very deep.

Table 8.1 lists the amount of carbon (in metric tons) that is fixed at three efficiency levels and five trophic levels. The tons of carbon are converted to tons of living organisms by multiplying by ten. Table 8.2 lists the average productivity and the total productivity, in terms of carbon fixation, of open ocean, coastal zones, and upwelling areas. As you can see, in the open ocean a mind-boggling amount of carbon is fixed in living organisms. Table 8.3 lists the carbon fixation, the trophic levels required, the growth efficiency, and the total fish production for each of the three ocean areas described in Table 8.2. More trophic levels are required in the open ocean because the primary organisms are much smaller there. In the coastal zones and upwelling areas the primary organisms are larger and tend to grow in clumps, and, hence, some carnivores can feed directly on them. Also, growth is much more efficient in these two areas. For these two reasons, a lower number of trophic levels are required in coastal zones and upwelling areas. In spite of the tremendous quantity of carbon fixed in the open ocean, only 1.6 million tons of fish are produced there. The other two areas provide 240 million tons. The open ocean is a biological desert as far as commercial fishing is concerned. Nearly all fish are produced in 10% of the ocean's waters.

The above calculations indicate that about 240 million tons of fish are

Table 8.2. Average and Total Production of Organic Carbon in Three Types of Ocean Waters

Area	% of Ocean	Area (million square kilometers)	Average Productivity (grams of carbon/m²/yr)	Total Productivity (million tons of carbon/yr)
Open ocean	90	326	50	16,300
Coastal zone	9.9	36	100	3,600
Upwelling areas	0.1	0.36	300	100
Total				20,000

Source: Ryther (1969).

produced yearly in the world's oceans. Yet production is not the same as harvest. Predators such as birds, tuna, squid, and sea lions probably consume as much as man harvests. The guano birds off the coast of Peru alone are estimated to consume 4 million tons of fish. Also, sufficient numbers of fish to reproduce the species must remain in the ocean. Predation and reproduction, then, account for an estimated 140 million tons of fish (see Table 8.4). This leaves an estimated 100 million tons for man to harvest. Fish landings in 1970 were an estimated 70 million tons, leaving a theoretical potential increase of only 30 million tons, which is unlikely to be realized with existing technology, since landings have been increasing very slowly since 1970.

The theory described above has been tested in two areas. One of them is the area between Hudson Canyon and the southern end of the Nova Scotian shelf, approximately 110,000 square miles. Theoretically, it should produce about 1 million tons of fish, which it did in fact in 1963, 1964, and 1965. Since then production has shown a decline that becomes more serious each year. The theory may be overly optimistic for this area, since the area was clearly overexploited in the 1960s and the 1970s. The catch quotas introduced under the 200 mile zone show promise of restoring the fishery to its former production.

Table 8.3. Estimated Fish Production in Three Ocean Areas

Area	Carbon Production (million tons)	Trophic Levels	Efficiency %	Fish Production (million tons)
Open ocean	16,300	5	10	1.6
Coastal zone	3,600	3	15	120
Upwelling areas	100	1½	20	120
Total				24×10^7

Source: Ryther (1969).

Table 8.4. Estimate of Fish Available to Man

Total fish production	240 million tons
Harvest by predators	70 million
Stock for reproduction	70 million
Available for harvest by man	100 million tons

Source: Ryther (1969).

The coastal upwelling area associated with the Peru Coastal Current is the most productive fishery in the world. This area, which comprises 24,000 square miles, should produce an estimated 20 million tons of fish annually, according to the theory. Assuming that this total is divided equally between man and predators, the 10 million tons available to man equals the actual landings in 1971. In December 1971, however, an ecological disaster overtook Peru: Anchovies disappeared from the fishery. The disappearance was attributed to the warm "El Nino" current displacing the cold, nutrient-rich Humboldt Current offshore.

This oceanic phenomenon had happened at least five times before, but never in association with a large fishery. The combination of the temperature change plus overfishing led to the collapse of the anchovy fishery (Fig. 8.2). The anchovy catch plummeted from 13,000,000 tons in 1970 to less than 200,000 tons in 1973, and continued downward. Quotas

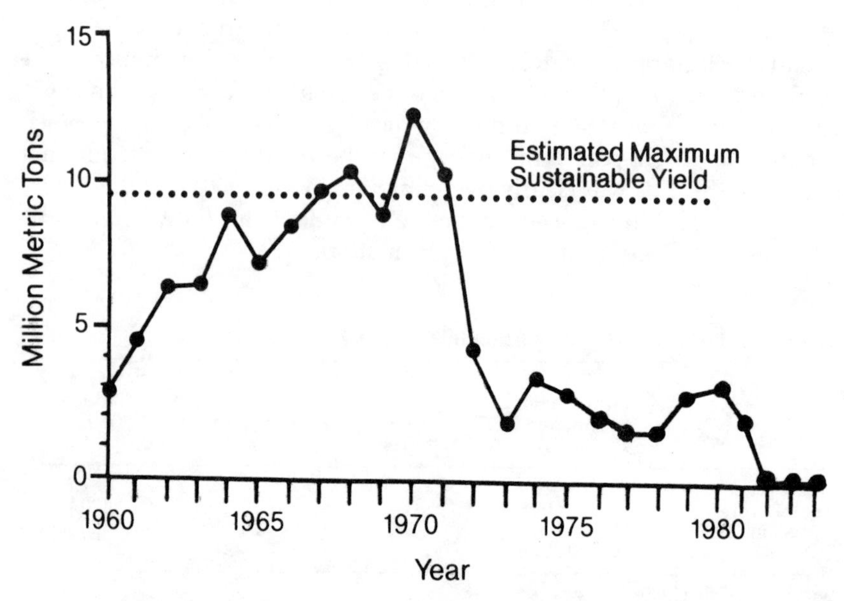

Fig. 8.2. Catch of anchovies off the coast of Peru.

on the fish catch were imposed in 1976, and a complete ban was established in 1980. There is some question as to whether the anchovy fishery will ever return, since sardines and mackeral are increasing. There are precedents for the disappearance of a specific fishery since it happened to the sardines off the coast of California and the pilchards off southwest Africa. There will be a fishery because of the favorable feed chain, but what it will be remains to be seen. But regardless of the reason for the disappearance of the anchovies, the Ryther theory may again have been optimistic in its prediction of the amount of fish an area should produce. The main conclusion to be drawn from these two examples is that the theory is probably over-optimistic, and this should lay to rest the hope of greatly expanded fish harvests.

One way to increase the take of "fish" from the sea is to harvest at a lower level on the production chain. As you can see from Table 8.1, the harvest of phytoplankton, for example, would be ten times as great as the harvest of herbivores. The Soviets and Japanese are attempting to do this by harvesting krill in the Antarctic. These small crustaceans, up to one inch in size, can be made into a paste that has an appetizing shrimplike flavor. Japanese experts have estimated that as much as 500 million tons of krill are available. Unfortunately, no technology is available to harvest krill economically. Krill form the basic diet of many whales in the Antarctic, and it might be more economical to allow the whales to harvest the krill and for humans to harvest the whales. This presupposes that the whale harvest is regulated such that an optimum, or at the very least, a replacement, number of whales is allowed to live, i.e., a "sustained yield" basis.

Present indications are that the number of whales is decreasing even though no species has become extinct. There is intense international pressure to reduce the whale harvest. The leading whaling nation is Japan, which was responsible for most of the total legal whale catch of 1400 in 1982. The International Whaling Commission in 1982 voted for a 5-year ban on whaling to start in 1986. Japan protested the ban. Under the Fishery Conservation and Management Act, the Federal Government can ban imports of fish and deny fishing rights within its 200-mile coastal zone to nations that do not comply with rules set by the International Whaling Commission. Since the value of Japanese fish catches and exports to the United States is much larger than the value derived from whaling, the choice would appear to be simple. Japan has just announced that they will honor a 5-year moratorium on whaling, starting in 1986.

It is likely that the moratorium on whaling will allow the depleted numbers of whales to increase. At that point, they can be placed on a sustained yield basis and an important resource will be maintained.

A fascinating way of harvesting krill in the Antarctic Ocean was proposed by scientists at the Northwest Fisheries Center, Seattle, Washington. They suggested that young salmon be released from hatcheries near the tip of South America into the currents around the South Pole. They would make one complete circuit around the pole in two years, or two in four years, and would then return to spawn in their release area. Determining the correct age of the fish at time of release and the correct date of release would be critical for the return of the fish. If the biological clock governing the urge to spawn gave the signal when the fish were 5,000 miles from the tip of Sooth America, the fish have no place to spawn and would be lost in the open sea. Large-scale hatcheries would be necessary, since southern South America has few rivers and therefore few natural spawning grounds. This would seem to be an excellent way of converting Antarctic krill into a highly desirable human food.

Some people believe that fishing technology still reflects the philosophy of the hunter. They compare the fishing industry with agriculture and note that hunting as a means of food production was abandoned thousands of years ago. We may eventually learn mariculture (salt-water farming), but research and technology are proceeding slowly. Mariculture still provides less than 10% of the marine food consumed in Japan, probably the most advanced nation in terms of mariculture. Food production by means of mariculture is very small worldwide. The production of fish by aquaculture (fresh-water farming) is well established but nevertheless constitutes only a very small percentage of the total fish harvest.

FISH PROTEIN CONCENTRATE

The dramatic increase in fish landings after World War II led to a surge of interest in fish protein concentrate (FPC). The need for greater supplies of animal protein was being voiced by the Food and Agriculture Organization of the United Nations and many other scientific groups. FPC seemed to be the answer. Fish were apparently available in unlimited quantities and could be made into a bland grayish powder with good nutritional properties that could be incorporated into soups, gruels, bread, noodles, and other foods. Almost every ethnic group consumed a form of food in which FPC could be utilized.

Research in FPC technology boomed after 1945, and over fifty methods of making FPC were proposed. Most of these processes involved solvent extraction, acid or alkali treatments, or mechanical approaches. The solvent-extraction approaches—which generally depended on two solvents, dichloroethane and isopropanol—were most successful. The dichloroethane process was developed by Ezra Levin. The process that probably has the most scientific support is the Halifax method. In this process, whole fish are ground up and the slurry is washed with acidified water. The mixture is then treated with isopropanol in countercurrent fashion. This means that the slurry moves in one direction and the isopropanol solvent moves in the opposite direction. The solvent removes oil, water, and some nitrogenous compounds from the mixture. The isopropanol is then removed from the mixture, and the final result is a practically tasteless creamy-colored powder with a very faint fish odor. Many variations of this product are possible, depending on the type of fish, degree of purity required, and so forth.

One of the entries (1970) into the FPC trade was the Astra Nutrition Development Corporation in Sweden, who teamed up with Nabisco, Inc. of the United States to make a high-quality FPC. The two groups intended to merchandize FPC in bakery products and perhaps also in the American school lunch program. They converted a 25,000-ton whaler into a floating factory to catch fish on the high seas and process them immediately into FPC. The process used only fish fillets, thereby ensuring a high-quality FPC with a low ash (that is, fluorine) content, and applied a modification of the isopropanol solvent-extraction method. Apparently, the venture was not successful.

FPC appeared to be a real panacea for the world protein shortage, particularly the shortage in developing nations. But for many reasons, it did not work. Perhaps the most important reason was the shortage of fish. The need for sophisticated technology, available only in countries that did not really need FPC, was another factor. The complicated problems of production, packaging, distribution, government regulations, and local preparation for consumption were not easy to solve. Not the least important reason was that the product could not be consumed alone; it had to be added to something, and this involved the complicated concept of consumer acceptance. Whatever the reasons, there is no appreciable production of FPC anywhere in the world today.

The history of FPC in the United States from a regulatory point of view is an interesting story. Serious research on FPC began in the U.S. Bureau of Commercial Fisheries in the Department of the Interior in 1961. In 1962, the Food and Drug Administration ruled informally that FPC could not be made from fish containing viscera or heads. The premise at that time was that no unwholesome material could be added to food, and surely intestinal contents were unwholesome. There may be some question as to the interpretation of this ruling, since the subsequent processing would certainly remove the unwholesome material.

However, the consensus at that time was that in order to be economical, FPC would have to be prepared from whole fish. The FDA decision was a major deterrent to the apparent incentive of American industry to invest time, effort, and money in FPC production. The Committee on Marine Protein Resource Development, appointed to study the situation, reported that FPC prepared from whole hake was safe, nutritious, and wholesome. In 1967, the FDA ruled that FPC could be made from whole hake as long as it contained 75% protein and as long as the protein was equivalent in quality to the protein in milk. A further regulation permitted domestic distribution only for household use and stipulated that packages of the product must weigh 1 pound or less. Bulk use of the additive by food processors would not be authorized unless preceded by the presentation of data demonstrating that the proposed use would not be deceptive to customers. The 1-pound regulation was another serious deterrent to the commercial development of FPC.

The first attempt to produce commercial quantities of FPC in the United States took the form of a $900,000 contract awarded by the Agency for International Development to Alpine Geophysical Corporation to produce 970 tons of FPC for Biafra and Chile. A plant was to be constructed by Ezra Levin in New Bedford, Mass. Winter storms, a shortage of hake, and technological problems prevented delivery of the FPC. Only 70 of the 265 tons delivered met the specifications, and the contract was cancelled.

The FPC program is a good example of an international cooperative food supply scheme which promised to be a panacea to cure world malnutrition. It was launched with great promotion and zeal but it never succeeded. The technology was sound, but it takes a good deal more than technology to guarantee success for international ventures.

ECOLOGICAL ISSUES

The sea and its inhabitants seem to have spawned more emotional ecological issues than any other part of our environment. Perhaps it is the degree of physical and intellectual development in some of these superb creatures. For example, the blue whale is the largest animal (130 tons) ever to inhabit the earth. It is larger than 30 elephants. The sperm whale (60 tons) has the most complex brain ever evolved. There is no doubt that their brains are used for intelligent complex communications. It is small wonder that they have evoked admiration for many conservation groups.

The fervor of some of the conservation groups has reached vigilante proportions, however. For example, Paul Watson, reinforced the bow of a converted steel trawler named the *Sea Shepherd,* with 8 tons of concrete and went after the infamous pirate whaler named the *Sierra.* Pirate whalers catch what they find, regardless of endangered species laws. Watson caught up to the *Sierra* off the coast of Portugal and rammed it. After the *Sierra* spent 8 months in port for repairs, someone attached a magnetic mine to its hull and the ship sank. The *Sea Shepherd* itself was captured by the Portugese navy but was scuttled before it could be auctioned off by Portugal. The *Sea Shepherd* alone cost $118,000, which was borne by the Fund for Animals headed by Cleveland Amory. Since the *Sierra* episode, two more pirate whalers have been blown up, and Watson has offered a $25,000 prize for anyone who sinks a pirate whaler.The existence of pirate whalers is inexcusable, but it is also a sad commentary that such a noble aim as preservation of the whales should inspire such violence.

The harp seals in the Gulf of Labrador off the east coast of Canada have evoked perhaps the most poignant responses. Harp seals in the first few weeks of their life are white and appear to be cuddly, lovable animals. Popular posters in the U.S. portray them as delightful creatures deserving of our utmost support. For many years, the young seals on the ice floes have been harvested by the natives of eastern Canada. Some of them derive as much as 40% of their income from the sale of the white furs. The popular method of slaughter is by clubbing followed immediately by skinning. Several conservation groups, notably the Greenpeace Foundation and the Royal Society for the Prevention of Cruelty to Animals considered this unacceptably cruel and mounted a public campaign. Protest ships positioned themselves between the hunters and the seals. Many young seals were sprayed by harmless red and green dyes as a visible protest. These protest movements had little effect, but another tactic did succeed. A movement spearheaded by the International Fund for Animal Welfare petitioned the European Economic Community to ban the import of white seal pelts. They were successful, and in 1983, the killing of young seals virtually ceased. A number of older seals will continue to be taken, but the numbers will be much smaller.

The conflict between the commercial interests and the environmental movements will continue, hopefully, as a trade-off. The seals in the Gulf af Labrador eat fish and, if allowed to reproduce at will, could affect commercial fishing. Therefore, their numbers will have to be controlled in some way.

The tuna–porpoise controversy is a vivid example of the food–environment problem. Prior to 1960, tuna were caught primarily by traditional hooks and poles, or by a technique called "long-lining." This method employs baited hooks at intervals on a main line a mile or more in length. In the 1960s a much more efficient technique was developed. Spotter planes would locate a school of tuna, and a net up to a mile in length would be set around the school. The planes located the tuna by looking for schools of porpoises, which, for unknown reasons, usually swim on the surface above the schools of yellowfin tuna. As the net is closed, the porpoises panic and, being air-breathing mammals, usually drown. The TV program called "Flipper" in the 1960's did more to raise public appreciation of porpoises than any other single event, and the protest movement gained momentum. It resulted in the passing of the Marine Mammal Protection Act, which called for reducing the number of porpoises killed. Porpoises and dolphins are very similar, and the terms are often interchanged, but porpoises are usually smaller. A "zero-kill" quota was suggested, but this caused turmoil in the tuna industry. All or part of the $500 million U.S. tuna industry threatened to move to other countries, namely Mexico. This would have been most unfortunate since other countries would not be bound by U.S. conservation laws, tuna and porpoises would continue to be taken together. Fortunately, cooler heads prevailed, and a quota of 20,500 porpoises was established. The fishermen modified their nets to allow most of the porpoises to escape prior to harvesting the tuna, and the porpoise kill was drastically reduced.

It seems that the food–environment conflicts will continue. The Argentine government has proposed harvesting 48,000 penguins in a trial program. The meat would go for human consumption and the skins for gloves and handbags. The decrease in predators has led to an increase in numbers of penguins, and since they eat fish, to a decrease in the fish harvest. The Association for the Defense of the Rights of Animals has already started a campaign.

BIBLIOGRAPHY

ROTHSCHILD, B. J., (editor) 1972. World Fisheries Policy. University of Washington Press, Seattle, Washington.

RYTHER, J. H. 1969. Photosynthesis and Fish Production in the Sea, *Science*, 3901, 72–76.

SHAPIRO, S., (editor) 1971. Our Changing Fisheries. U.S. Govt. Printing Office, Washington, DC.

9

Potential Sources of Protein

The world's population currently derives about 70% of its protein from plant sources. It is unlikely that the proportion of protein from animal sources will increase in the foreseeable future, in view of the increased pressure on the land for total food production.

It may well be that the future will see more emphasis on total food production and less on the relative importance of protein. However, the past generation has seen great importance attached to improved protein supplies, and hundreds of new protein formulations have been developed for infant and child feeding. This emphasis has stimulated much more research and development in the area of protein than in the corresponding area of carbohydrates and fats. In spite of this research, protein in the human diet will probably continue to be derived primarily from conventional plant and animal sources. A description of these phases of agriculture is beyond the scope of this chapter. Our purpose here is to suggest that there may be something additional, and hopefully better, on the horizon.

PROTEIN FROM SOYBEANS

Because it has been publicized as a possible replacement for animal protein, soy protein has captured the imagination of the American public. Americans eat too much meat, we are told, and a partial switch to plant protein would be beneficial. We will probably see a gradual switch to plant proteins for economic reasons if nothing else.

Soy products have traditionally been associated with oriental cooking but in recent years America has dominated the soybean market. In 1982, the world production of soy beans was about 87 million tons, of which the United States produced 62%, Brazil 15%, and China 9%.

Soybeans were practically unknown in the United States in 1925, but they rapidly became the second largest plant crop in the United States. Today, they are second only to corn in importance. The reason for this amazing increase in production is that soybeans are a good source of both edible oil and protein. In 1940, in the United States, seven times as much butter, as compared with margarine, was consumed. By 1972, margarine was in the lead by two to one, and approximately 83% of American margarine is made with soy oil. In addition, soy oil comprises 90% of prepared dressings, 80% of salad and cooking oils, and 62% of the shortenings. Soy oil completely dominates the American edible oil industry. The protein portion of the soybean has become the mainstay of the meat and poultry industries. It is the major component of the high-protein animal rations.

It should come as no surprise that with such a large investment in soy protein in the animal industries, major efforts were made to introduce more soy products into the human food chain. A number of products are available today, and their consumption is rising slowly.

Food-Grade Soy Products

Food-grade soy products apart from oil can be conveniently divided into four areas: whole beans, flours, concentrates, and isolates.

Whole Beans

These are used in the United States in very small amounts. They are canned in the immature stage as green beans in the pods similar to conventional green beans. The mature beans are also canned in tomato sauce or ground with water to make a milklike beverage. Such beverages have been very successful in other areas of the world, but not in the United States. A number of soy-based beverages are sold as formulas for infants allergic to cow's milk, but the volume of such beverages is still small.

A considerable portion of the soybean crop is used to make traditional oriental seasonings and high protein foods. Soy sauce is made in the Orient by mixing cooked defatted soy flakes with wheat, inoculating this mixture with *Aspergillus oryzae*, and salting and fermenting it for 8–12 months. In the United States, soy sauce is usually made by acid hydrolysis. *Tofu* is made by precipitating the curd from soy milk. *Tempeh* is made from cooked soybeans inoculated with *Rhizopus oligospores* and fermented for 24 hours; the resulting cakelike mass is sliced and fried. There are many other soybean-based foods popular in the Orient, such as dried *tofu*, *kinako*, *miso*, and *natto*.

Soy Flours

Soy flours and grits are important but relatively unknown items in the United States because they are usually incorporated into other products. The only difference between flours and grits is particle size; flours are ground finer. Both are made in a defatted form, a full-fat form, and any combination in between. Full-fat flour is made by cleaning the soybeans, cracking them to remove their hulls, heating them to minimize their beany flavor and to inactivate their enzymes, and drying and grinding them. Defatted flour is prepared in a similar manner except that after dehulling and crushing, a solvent extraction process—usually with hexane—is introduced to remove the fat. A defatted flour is approximately 51% protein, 1.5% fat, and 34% carbohydrate. A full-fat flour is 41%, 21%, and 25%, respectively. An estimated billion pounds of soy flour were produced in 1976.

Protein Concentrates

Protein concentrates are prepared by removing the water-soluble sugars, ash, and other minor low-molecular-weight components from defatted soy flakes. These components are usually extracted with either water or alcohol. The protein content of the resulting product is usually 66–70%. The consumption of protein concentrates in the United States in 1976 was estimated at 84 million pounds.

Protein Isolates

These are prepared from defatted soy flour in essentially two steps. Flour is treated with dilute alkali so that the mixture has a pH of 7–9 and the protein dissolves. The insoluble residues are removed, and the liquid is acidified to a pH of 4.5. The protein precipitates and is filtered off as a curd. The protein can be redissolved, and extruded in any shape or passed through fine nozzles to produce threads that are not unlike nylon. Actually, the protein threads are produced on equipment very similar to that used to produce nylon threads. A dried bundle of soy isolate threads looks like blond hair. The threads can be bundled together, colored, flavored, and packaged to simulate chicken, ham, scallops, beef, and other foods. The protein content of a typical isolate is 93–95%. The American consumption of protein isolates in 1976 was estimated at 87 million pounds.

Uses of Soy Protein

Protein that is incorporated into other foods is generally classified as "functional" or "filler" protein. Filler protein contributes to the protein content of the food, and that is all. Functional protein also contributes to the protein content, but in addition it has some functional value, such as emulsification, fat absorption, water absorption, or texture.

Emulsification

Emulsification is a very important property of soy proteins, which can stabilize both oil-in-water and water-in-oil emulsions. Soy flours, concentrates, and isolates are used extensively in ground meat products such as sausages and wieners in order to make them more stable products. They are also used in baked goods and creamed soups for the same reason. In addition, they can be incorporated as effective emulsifiers in whipped toppings, frozen desserts, simulated ice creams, and confections.

Fat Absorption

Soy proteins have the ability to bind fat in products such as hamburger and sausages, thereby reducing the fat that is lost in the cooking process. The flavors are retained in the fat portion; thus, the products are usually juicier and more flavorful. With products such as pancakes and doughnuts, the soy protein decreases fat absorption. The actual phenomena involved in the ability to decrease fat loss and minimize fat absorption are not well understood at this time.

Water Absorption

Soy proteins absorb water and tend to retain it in finished food products. This is a very useful function in baked goods, pet foods, and simulated meats. The addition of soy protein to macaroni, however, decreases water absorption, thereby producing a firmer, more desirable product. Adding soy protein to hamburger patties decreases their drip loss and produces a juicier product.

Texture

Soy proteins can provide desirable textures to many products. The simplest example of this is the thickening of soups and gravies. The gelling properties of soy protein in sausages and luncheon meats is more complicated but nevertheless effective. The tendency of soy protein to gell upon heating is utilized when, for example, pet foods are put into a can in the form of a slurry. The subsequent heat in the canning process imparts the desired texture to the product.

Functional protein in foods has many other uses, such as cohesion, elasticity, film formation, and color-control aeration. With these obvious advantages, soy proteins are being added to many foods. Soy proteins are also being added as filler proteins to a number of products in order to increase their protein content. They can be added conveniently to almost any homogenized product, such as hamburger, luncheon meats, and chili. They can form up to 30% of the protein in the school-lunch program.

Conclusions

The major uses of soy proteins in the human food supply are in the form of flours, grits, and concentrates. These are used for functional reasons and constitute up to 3% of the food to which they are added. The use of isolates in the form of textured products has received most of the publicity connected with the replacement of meat by soy products. However, the manufacture of textured soy products requires considerable sophisticated processing. Consequently, a simulated chicken meat, for example, is relatively expensive. Such products should enjoy a cost advantage over traditional meat, but the price differential and the ultimate quality of the products will probably determine their degree of success in the marketplace.

Soybans are never eaten raw. All of the recipes and processes employ either a heat treatment, a fermentation process, or a physical process to purify the product. There are good reasons for this. Raw soybeans contain at least 12 toxic components, which have to be broken down or removed prior to human consumption. Scientists have known for 40 years that something in soybeans made them difficult to digest. One of the compounds inhibits the trypsin enzyme, which is produced in the pancreas and is essential to digestion. Plant breeders in 1979 found a way to produce soybeans without this inhibitor (the Kunitz trypsin inhibitor) and commercial varieties are expected within the decade. There are other causes for concern (see Chapter 2).

PROTEIN FROM PETROLEUM

The prospect of creating protein by growing microorganisms on petroleum products has captured the imagination of researchers in the past fifteen years. The prospect of obtaining high-quality protein from a factory instead of an animal farm is exciting indeed, and a great deal of research and development has gone into this area. It is one of the most promising ideas for food production available at this time.

The concept of making usable protein from microorganisms is not new, but large-scale commercial trials have been attempted only in the past 15 years. The researchers at the Massachusetts Institute of Technology coined the term "single cell protein" (SCP) as a general term for protein from microorganisms.

The prospect of making protein from petroleum started with the observation that yeasts were growing in the wells being drilled for crude oil. It was obvious that the yeasts were using petroleum as a source of energy. Actually, it had been known even before recorded history that something would grow on fruits and vegetables and form tasty beverages, and the residue could be used as food for animals. They did not know that it was yeast and alcohol, but they knew they liked it. The residues from the fermentation processes contained additional protein from the bodies of the yeasts. However, what really fired the imagination was the possibility of using a very cheap (in those days) substrate (crude oil) in a very economical way. One hundred pounds of hydrocarbons from oil plus 200 pounds of oxygen from the air would produce 100 pounds of cells. Compare this to the need for 200 pounds of carbohydrates from sugar plus 69 pounds of oxygen to produce 100 pounds of cells. The reason for this is that oil is a more concentrated source of carbon. Also, this discovery was made before the rise in oil prices. Actually, almost any source of carbon can be used as a substrate for growing microorganisms. The concept is very simple, as illustrated in Fig. 9.1. A substrate with nutrients, water, and oxygen from the air is pumped into a vertical tube called a fermenter, and the output is drawn off from the top. The output consists of cells, other chemicals called metabolites produced in the fermentation step, carbon dioxide, and heat. The choice of a substrate is determined by cost, availability, safety, ease of purification of the cells viscosity, and many other factors.

The British Petroleum Co. decided to exploit the possibility of producing protein from petroleum. In 1970, they built a 4000-ton-per-year pilot plant in Grangemouth, Scotland. This plant used specially purified C-10 to C-18 alkane hydrocarbons from petroleum. They also built a 20,000-ton-per-year plant in Lavera, France designed to use a portion of the distillate from the petroleum refineries. The yeast used the hydrocarbon as a source of carbon for energy, and what they did not use was merely sent back to the refinery. Both plants used a yeast called *Candida lipolytica*.

The two pilot plants were adequate to provide SCP for feeding trials with a variety of animals. Protocols for testing SCP for both animal and human food uses were established by the Protein Advisory Group of the World Health Organization–Food and Agriculture Organization. The

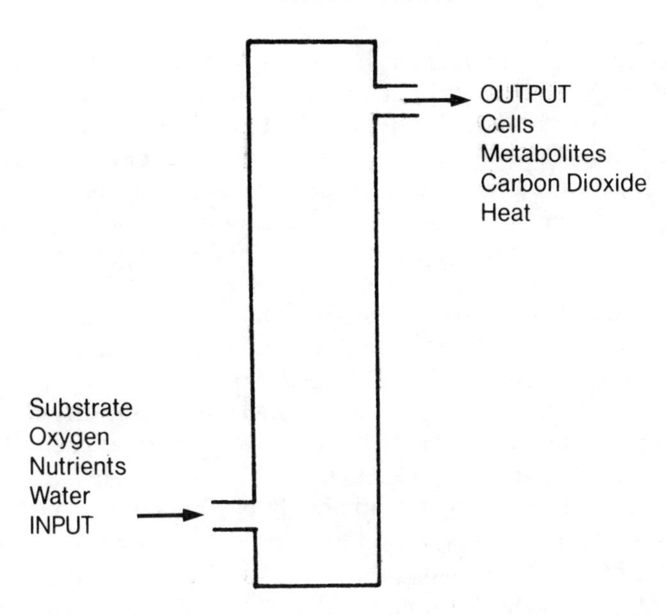

Fig. 9.1. Fermenter to produce microorganisms from an organic substrate.

safety tests alone involved multigeneration feeding trials with rats, dogs, and at least one other species for periods up to 2 years. Economic and acceptance studies on chickens, pigs, cattle, and other animals took about 6 years. British Petroleum provided the required data on safety, acceptance, supply, economics, etc., and the stage was set for commercial exploitation of SCP. It is easy to understand the European desire for a second source of protein concentrates for animal feed after the disappearance of Peruvian fish meal left them with American soy meal as the sole source. Eight members of the European Economic Community, Belgium, Denmark, France, West Germany, Iceland, Luxemburg, the Netherlands and Great Britain, all approved the use of SCP for animal feed.

British Petroleum teamed up with Anic, the petrochemical arm of the Italian state-owned hydrocarbon company to build a plant in Sardinia. Another Italian company, Liquichemica built a plant in Calabria. Both plants employed purified hydrocarbons and had an annual SCP capacity each of about 100,000 metric tons. Both plants obtained construction permits from the Italian government and construction was completed in 1976. Both plants then obtained a "shake-down" permit for trial runs and then applied for an operating permit. They were refused. Two reasons were given by the Italian health authorities. One was that the plants produced too much dust. However, a cement plant up the coast produced 40 times as much dust and the prevailing winds usually blow off-shore. The second reason was that about 70 ppm of waxes were found in the fat of pigs fed Toprina (the trade name for BP product.) For comparison, the American FDA allows 950 ppm in meats and as much as 1500 ppm in breads. The reasons for the permit refusal is hard to understand. At any rate, the cost of $20,000,000 per year to keep the plants viable, but idle, was too much, and both plants were sold for scrap in 1978. British Petroleum withdrew from the SCP business. In the meantime the potential for SCP production was being explored in Saudi Arabia and Venezuela. Japan had three large commercial plants in the blue print stage, but some environmental groups raised the spectre of possible minor contamination of the SCP with benzopyrene, a known carcinogen. Benzopyrene is found in trace amounts everywhere in crude oil, coal, and many other hydrocarbons. Public pressure forced the cancellation of the plans in Japan and the oil-producing countries. That was the end of the European plans to produce SCP from petroleum.

The scientists at Imperial Chemical Industries (ICI) in England undertook the development of the second-generation plants for SCP. They had ample quantities of natural gas from the North Sea fields and attempted to grow microorganisms using natural gas as a substrate. But the organisms grew poorly. It was discovered that the organisms were making methanol from the natural gas first, so the substrate was changed to methanol instead of natural gas. ICI already knew how to make methanol from petroleum; they are world leaders in this technology.

ICI built a plant at Billingham, England to produce 110,000 metric tons of SCP per year. It came into full production in 1980. The procedure uses the bacteria *Methylophilus methylotrophus.* The product is sold under the trade name "Pruteen." The major marketing thrust appears to be directed toward animal feed.

In Germany, similar efforts were being made to develop SCP. The German chemical giant, Hoechst, developed a process to grow a bacteria, *Methylomonas clara,* on methanol and built a 1000-ton-per-year pilot plant in 1978. Their product is being sold under the trade name "Probion." They have developed processes to lower the content of nucleic acids in the protein product to make it suitable for human use. The consumption of nucleic acids in amounts over 2 grams per day is not recommended for humans because of the tendency to produce gout, which is a very painful disease caused by precipitation of uric acid crystals in the joints. Hoechst has announced its plans to pursue the development of SCP for both animal and human consumption.

In the United States, the American Oil Company developed a process to grow a Torula yeast on ethanol. In 1975, they built a 5000-ton-per-year plant at Hutchinson, Illinois using ethanol from petroleum. Ethanol is a more expensive substrate than methanol, but it is more acceptable to the public than methanol. Regardless, the company is using a different marketing approach. Their product, "Temptein," is being marketed as a flavor adjunct, i.e., as a functional rather than a filler protein.

Because the Soviet Union is a large exporter of petroleum, it would seem natural for them to produce protein from petroleum. The 200-mile economic zone limited the fish catch for the Russians, obviously decreasing their source of protein. Their climate is too cold for soybeans, so their major source of protein concentrates for feed is sunflower seed meal. A plant to produce protein from purified petroleum waxes using a yeast was built at Gorky, and apparently it is the only plant in the world using petroleum. They have expressed an interest in using the technology developed by ICI.

There is no question that protein from microorganisms will become much more important in the near future. The stakes are high. It will first be introduced as a major animal food source and later as human food.The market for protein concentrates is estimated at up to 2 million tons, and soy meal is obviously supplying the major share. SCP is not subject to the vagaries of nature and, if the economics are promising, it may well provide a large share of the protein for animal feed. Having two major sources of supply is standard practice for most industrial concerns, and ensures continuity of supply for the users.

PROTEIN FROM CARBOHYDRATES

The production of SCP from yeast grown on carbohydrates is a major industry around the world. Nearly all of the SCP produced in this way is used in the animal feed industry, but small quantities are incorporated into the human food chain. Three major types of SCP are produced from yeast: molasses, spent sulfite liquor, and whey.

The production of SCP from molasses, using the yeast *Saccharomyces cereviseae,* is the largest of the three types. Beet-sugar molasses is preferred to cane-sugar molasses because it contains more nutrients, but both are used in large quantities. Conventional fermentation in large tanks optimized for the growth of yeast rather than the production of alcohol is used. The product is called Primary Dried Yeast.

SCP may also be produced from spent sulfite liquor from the pulp and paper industries. The composition of the sulfite liquor varies with the source of wood and type of process but may average 2–3% sugar. The sugars are fermented by the yeast *Candida utilis,* which is then recovered as a source of SCP. Considerable sulfite liquor is being dumped at present, and since it is a source of pollution it may become more economical to use it as a source of food for yeast. Yeast from this source is called Torula Dried Yeast.

SCP may also be produced from whey, a by-product of the production of cheese. The lactose sugar in whey can be fermented by the yeast *Saccharomyces fragilis,* which produces a product called Dried Fragiles Yeast. Much of the whey currently being produced is dumped, again creating a pollution problem. Considerable research has been devoted recently to methods of treating whey to recover high quality milk proteins present in low concentrations in the whey and to use the sugar for fermentation. The problem with this concept is that the cheese plants are usually small and scattered, making collection a problem for large-scale yeast plants. Since less than half of the available whey is now being recovered, there is no doubt that both production of yeast protein and milk protein from this source will increase.

Brewers' yeast is another potential source of SCP. An estimated 25,000 tons of dry brewers' yeast could be available as a by-product of brewing 100 million barrels of beer. Less than half of this is currently being recovered.

The production of SCP from yeast by fermentation could be very large if all the available existing raw materials were used. For example, an estimated 1,000,000 tons could be made from molasses, 500,000 tons from sulfite liquor, and 400,000 tons from whey. Obviously, these figures are not likely to be realized fully, but the potential is large.

PROTEIN FROM LEAVES

Among novel sources of protein, green leaves have the most potential for production. However, the technology of leaf protein concentrate (LPC) has been slow to develop, and production at the present time is very small.

The basic technology for extracting protein from leaves was developed in Europe, particularly Hungary, in the 1930s and was pursued vigorously in England by N. W. Pirie. The concept of protein production from leaves received considerable support in England in the early 1940s as a result of the possibility of a food supply blockade during World War II. The International Biological Program from 1964 to 1974 also provided considerable encouragement.

The techniques developed by Dr. Pirie and his supporters were deliberately simple. The assumption was that LPC was needed more in the developing countries, and that sophisticated technology would therefore be inappropriate. The LPC program was divided into production, processing, nutrition, and acceptability. Many types of plants were tested for their ability to produce protein under a wide variety of horticultural and agricultural conditions. Suitable plants were chosen for processing studies.

The plants are pulped and pressed in one operation that separates the fibrous matter from the green slurry. The green slurry containing the protein is heat-treated, which coagulates the protein, and is then pressed. The final product is a green curd, which can be dried and added to other food products. The curd contains approximately 40% protein of good biological value and has become a successful high-protein component of animal feed.

Unfortunately, for human purposes, the green crude protein has a bitter taste and an unattractive color. Dr. Pirie has developed a series of human food dishes, such as soups and baked goods, that use LPC. It is obvious that they all have to be highly flavored and very dark in color. A safe conclusion is that they have not been particularly well received.

Because the dark green, bitter protein preparation has limited appeal, a new approach has been to purify the protein. One example, developed at the USDA laboratories, uses alfalfa leaves. Alfalfa was chosen because alfalfa meal was already well established in the United States primarily as a source of yellow pigment for poultry. In this process, fresh leaves are ground and pressed. The liquid is heated to 60°C; this coagulates the chloroplast proteins, which are filtered off as a green curd. The liquid is then heated to 80°C; this coagulates the cytoplasm proteins, which are filtered off as a gray curd. The resulting liquid, known as "alfalfa solubles," can be dried and, together with the original press cake, be sold as animal feed. The protein curd resulting from both pressings is obviously rich in protein and has been developed as a feed for nonruminant animals. The next phase is to purify the protein concentrate from the second pressing even further, in order to create a source of protein for humans. The process described above was commercialized under the trade name Pro-Xan.

Another purification added to the processes described above would introduce a degree of sophistication that might be inappropriate for developing countries, but such a purification would be essential for human acceptability in the United States. Harvesting aquatic weeds, now considered a major nuisance in the South, and processing them as a source of protein is a fascinating idea, but the economics are something else again. In spite of its vast potential, LPC is unlikely to make much of an impact as human food in the near future. There may be one important exception. In countries like India where vegetarian diets are common, many foods are greenish brown in color and very highly spiced. Neither the green color nor the bitter taste of LPC would be a disadvantage with these types of foods.

PROTEIN FROM ALGAE

As a source of SCP, algae (*Chlorella*) has received considerable publicity in the press and in the scientific community as another "factory" for protein. Algae can be grown from a very simple inorganic source of minerals plus carbon dioxide from the air and sunlight. Many countries have grown algae in large-scale laboratory setups, but the need for sunlight as an energy source has limited the development of large plants. The use of artificial light in a "factory" to grow algae is far too expensive.

An interesting development in pollution control may make the production of algae economically feasible. In California, considerable research has been done to optimize the growth of algae so that it produces oxygen and removes minerals from sewage effluents. In effect, the algae are being used to clean up the effluent from sewage and other waste disposal plants before this liquid is released into rivers. Ponds 10 feet wide, 6 inches deep, and hundreds of yards long are employed for this purpose, and they are very effective. Sewage is an excellent source of nutrients for algae, and it is one commodity in plentiful supply. The algae are at present being produced as a by-product and are not being utilized. Apparently, the production of SCP from algae is uneconomical at the present time due to the cost of recovery and purification. If the economics change, SCP from algae may be available in large quantities.

Consumption of algae by humans is not new. The Chinese and Japanese have eaten algae in the form of seaweed (*Porphyra, Chondrus,* and several other varieties) for centuries. The natives around Lake Chad in Africa have eaten *Spirulina maxima* since ancient times. *Spirulina maxima* is also produced in large quantities in Lake Texcoco in Mexico. The ocean has been estimated to produce a mind-boggling 20 trillion tons of algae annually—more than 5 tons for each person on the earth today. Unfortunately, the maximum concentration of algae in seawater is about 3 mg per liter, and the minimum considered feasible for harvesting is about 250 mg per liter. Also, some algae are toxic to humans. The red tide (*Gonyaulax*), which periodically causes large fish kills on the Atlantic coast, is a good example. There is currently no known technology for the economical harvesting of algae from seawater. We will have to let the fishes do it for us.

PROTEIN FROM FUNGI

We do not normally consider fungi a source of protein, and probably rightly so. For example, the 120,000 tons of mushrooms consumed yearly in the United States provide about 0.05% of the required protein. Fungi are important in the United States, but primarily as flavor enhancers in cheese and condiments. This is not the case in the Orient, where large quantities of fermented products from soybeans, wheat, rice, copra, peanuts, fish, and other foods are produced by inoculation with fungi. Fungi do increase the protein content of foods, particularly high-carbohydrate foods. It is possible to exploit this capability with almost any type of carbohydrate food. For example, the protein content of rice can be doubled by inoculation with a fungus (*Trichoderma* sp.). Manioc roots (cassava or tapioca) have a protein content of about 0.7%; after inoculation with a fungus and 4 days growth, this figure can be increased to 5.7%. Sugar cane is probably the most efficient user of sunlight in the plant kingdom. Its photosynthetic apparatus is capable of producing large quantities of carbohydrates (sugar), yet, ironically, some strains actually produce more protein per amount of land than soybeans. The protein in sugar cane is so dilute as to be unusable. However, the carbohydrate in the cane can be the substrate for a fungus capable of producing nearly 3 tons of protein, compared with the 800 pounds produced by an acre of soybeans. White potatoes, sweet potatoes, corn, sorghum, millet, and many other foods are good candidates for this method of increasing a food's protein content. W. D. Gray (1970) has proposed that "vastly greater contributions to the world protein supply could be made if the carbohydrate-synthesizing capabilities of the green plants were combined with the protein-synthesizing capabilities of the non-green plants."

Undoubtedly, protein supply can be increased by the above methods, but there is much to be done before the resulting products are likely to be accepted by the Western world. We are not used to eating fungi, except in the form of mushrooms and in cheese, so the flavors would be quite different. There is also a problem of contamination with fungi, which produce toxins. However, existing technology can handle both of these problems.

MISCELLANEOUS SOURCES

Many suggestions have been proposed for the production of proteins for human consumption. For example, cottonseed meal contains high-quality protein, yet it has not been used to any extent in the human diet. The reason for this is that cottonseeds contain a toxic yellow pigment called gossypol. The heat treatment required to degrade the gossypol lowers the quality of the protein. Nearly all the cottonseed meal currently available is fed to cattle, who apparently can tolerate the gossypol content. Animals with one stomach are fed the heat-treated meal. There are two potential solutions to the gossypol problem. First, a new variety of cotton that contains little or no gossypol in its seeds has been developed. Unfortunately, the cotton fibers in this variety are not quite as good as those in the regular varieties, so this important development needs some refining. Second, the USDA laboratories have developed a process (the liquid cyclone process) that removes the pigment glands that contain the gossypol. The final flour contains about 65% protein of good biological value. A pilot plant producing 25 tons per day is now in operation. The potential of protein from oil seed is very large, since the current world-wide production of cottonseeds is estimated to be over 100 million tons.

Protein from peanut meal is a good possibility in countries such as India, where peanut meal is in good supply but is used mainly for fertilizers. Animal protein from herds of antelope-like animals in Africa is said to be underutilized. We should develop a type of animal husbandry more suited to them. Insects have been used as a source of food for man for a long time. The giant African snail, a creature about 8 inches long, is reputed to be quite tasty and a good source of edible protein. Fried termites, caterpillars, ants, rats, lizards, locusts, bird nests, and, indeed, almost every conceivable type of organic matter containing protein has been utilized at some time by some culture as a source of protein. Our only comment on some of these is that most people are not quite ready, or hungry, for them yet.

BIBLIOGRAPHY

GRAY, W. D. 1970. The Use of Fungi as Food and in Food Processing. Chemical Rubber Co., Cleveland, Ohio

KUMAR, H. D., and SINGH, H. N. 1971. A Textbook on Algae. Affiliated East-West Press, New Delhi, India

MATELES, R. I., and TANNENBAUM, S. R. 1968. Single Cell Protein. The M. I. T. Press, Cambridge, Massachusetts

PIRIE, N. W. 1971. Leaf Protein: Its Agronomy, Preparation, Quality and Use. IBP Handbook No. 20. Blackwell Scientific Publications, Oxford, England

DE PONTANEL, G. 1972. Proteins from Hydrocarbons. Academic Press, New York

TELEK, L., and GRAHAM, H. D. 1983. Leaf Protein Concentrates. AVI Publishing Co., Westport, Connecticut

WOLF, W. J., and COWAN, J. C. 1971. Soybeans as a Food Source. CRC Press, Cleveland, Ohio

10

Improvement of Food by Nutrification

In earlier generations, mankind relied on a varied diet to obtain an optimum nutritional state. This is still a worthy goal, but for several reasons it may not always be feasible. Pressure on the land to produce calories may reduce the choices available in some of the more densely populated countries. In some of the affluent countries, people may choose not to eat a varied diet for religious or personal reasons. In any case, it is of overriding importance that we make the existing food as nutritious as possible. Nature did not make a wheat kernel to nourish man; the primary purpose of the kernel is to create another wheat plant. The nutrients in the kernel necessary to produce a wheat plant are not necessarily the substances that produce optimum human nutrition. It is up to human ingenuity to discover the combinations of wheat kernels and other foods or synthetic nutrients that provide optimum nutrition. This is true for all staple foods.

COMBINATIONS OF FOODS

Combinations of foods for optimum nutrition have been known for thousands of years. Trial and error has probably resulted in desirable combinations of food in every culture. The rice-eating nations combined rice and fish as a source of protein. The fermented fish dishes such as *nuoc mam* and soy sauce in many Asiatic cultures added protein to a largely carbohydrate diet. The addition of milk to oatmeal and cheese to apple pie had the same effect. The combinations of vegetables, starchy roots, and meat in many cultures were balanced for protein and carbohydrates. In modern times the concept of blending one source of food with another has received great impetus from the development of balanced high-protein formulations for infants and children. It is a short step from this to balance the diet with protein, carbohydrates, fat, minerals, and vitamins from many sources. The United States at least is moving rapidly in this direction.

ADDITION OF VITAMINS AND MINERALS

The addition of nutrients to the diet involves four terms:

1. *Enrichment:* addition of one or more nutrients (naturally present in the food in lesser amounts) in order to increase consumption of these nutrients.
2. *Restoration:* addition of nutrients to a processed food to replace nutrients lost during processing.
3. *Fortification:* addition of nutrients that may or may not be naturally present in the food in order to increase consumption of those nutrients by the general population or a segment of the population.
4. *Nutrification:* a general term for the addition of nutrients to food.

The first three are legal terms as clarified in the proposed FDA regulations on nutrition-quality guidelines.

The addition of vitamins to foods became possible after the exciting research in the 1930s, during which nearly all of the vitamins were discovered and synthesized. Today, all of the vitamins necessary for human nutrition can be synthesized except vitamin B_{12}, which has a particularly complex chemical structure. Fortunately, this vitamin can be extracted as an inexpensive by-product of the mold used in the preparation of the antibiotic streptomycin. Most of the vitamins can be produced very inexpensively. For example, vitamin C costs less than 1 cent a gram. This paved the way for the addition of vitamins to many foods. Since minerals have always been relatively cheap, they were also added as human biochemical needs became evident.

Food fortification probably started in 1833 when the French chemist Boussingault suggested that iodine be added to table salt. Iodization of table salt was encouraged in Europe in 1900 following the discovery of thyroxine by Bauman in 1895. Iodized salt became available in the United States in 1915. Vitamin A was added to margarine in Denmark in 1918, and margarine, fortified with vitamin A and sometimes vitamin D, became available in the United States in 1931. The discovery by Steenbock in 1924 that vitamin D could be made by irradiating ergosterol from yeast was a nutritional milestone. It made the foul-smelling and even worse tasting cod liver oil obsolete. It became feasible to add vitamin D to milk, which already had a good supply of calcium and phosphorus. This one development practically eliminated rickets as a childhood disease. The addition of vitamin D and vitamin A to milk became widespread in the 1930s and 1940s. In the United States, addition of thiamin, riboflavin, calcium, and sometimes iron to white bread became an established procedure in the 1930s. In 1941, legislation in the United States *required* the addition of thiamin, riboflavin, niacin, and iron to white flour and bread.

The addition of vitamins to food has been a very successful concept. Perhaps two examples will dramatize the early successes: (1) the addition of niacin to cereals in the U.S. virtually eliminated pellagra (Fig. 10.1) and (2) the introduction of enriched rice greatly reduced the incidence in Japan of diseases due to lack of primarily three vitamins, thiamin, riboflavin and niacin (Fig. 10.2).

The low cost and easy availability of vitamins has led to another development, one that may not be quite as desirable. The widespread promotion of vitamin pills and vitamin and mineral tonics became a fixture of American life. Undoubtedly, they have helped the nutritional status of the nation, but their promotion may have been overdone. Many people consume more vitamins then they can possibly use. Except in the case of vitamins A and D, which can be toxic if taken in excess, probably the only harm involved is an economic loss. Perhaps even this is compen-

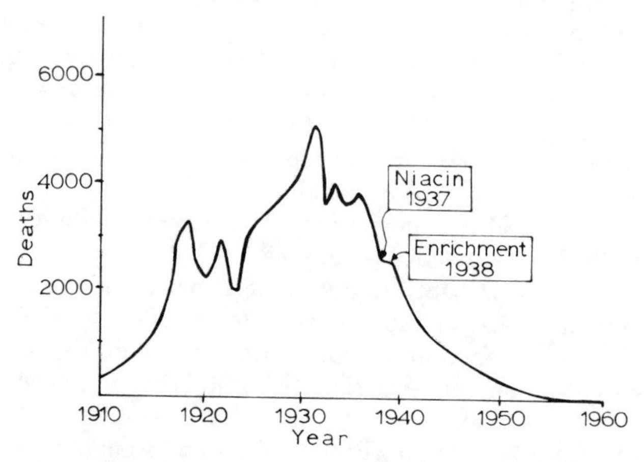

Fig. 10.1. The effect of addition of niacin to cereals on the incidence of pellagra in the United States.

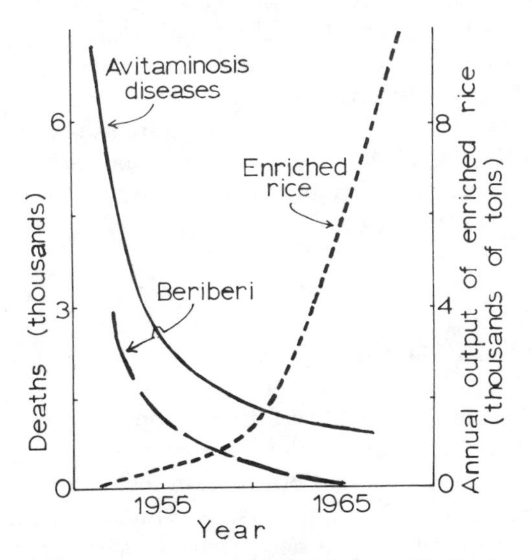

Fig. 10.2. The effect of the enrichment of rice with B vitamins on the incidence in Japan of related diseases.

sated for by the psychological importance of thinking that one is actually doing something for one's health.

The availability of low-cost vitamins and minerals has led to the development of a large number of specialized formulations designed primarily for infants. The social movement against breast feeding in past generations led to a sizable market for milk substitutes. These were balanced for protein, carbohydrate, fat, mineral, and vitamin content according to the best nutritional information available. Some experts in infant nutrition think that the complexity of the formulation and the modification of the nutrients has perhaps been overdone, but on balance the infants have unquestionably been better nourished. The variety of infant formulations may be startling to the uninitiated, but compared with the infant diets available in the 1920s they are obviously a vast improvement. One can obviously make a case for the nutritional and health advantages of breast feeding, but there will always be a place for human milk substitutes as there will for prepared baby food.

The nutrification of baby food led to the next logical step in nutrification—formulated complete breakfasts for calorie-conscious adults. By drinking one can of liquid, one can be assured of a food intake of known calorie content and a balanced content of all required minerals and vitamins. The protein and fat content can be adjusted to any desired level and, in the case of fat, to any desired ratio of saturated to unsaturated. Another example of this type of formulation is the cakelike preparations that can be consumed at any meal. They have the same nutritional characteristics as the liquid breakfast preparations. The cakelike preparations have not yet received widespread acceptance, perhaps because their satiety value may be too high; one just cannot eat much of them. It is too early to assess their acceptance yet, but they may be a preview of things to come.

ADDITION OF PROTEIN

The nutrification of food by the addition of protein received considerable impetus from the intense effort to develop and introduce high-protein foods for infant and child nutrition in the era from 1950 to 1970. Protein malnutrition had been touted as the primary nutritional problem in many of the developing countries, particularly those in the warmer climates. Efforts to develop high-protein foods were not aimed at the adult population, since it was obvious that many adults ate cereal grains with practically no extra source of protein and were healthy. An adult can subsist on the protein obtained from an adequate supply of cereal grains, but infants and children cannot. They need food with a higher ratio of protein to calories in order to enjoy optimum growth. This is not to say that adults cannot also profit from protein beyond that obtained from cereal grains. Their nutritional state will be better for it.

The improvement of protein in cereal foods can be accomplished in several ways. One can add synthetic amino acids to increase the amount of the essential amino acids that are limiting in a particular cereal. An amino acid is said to be limiting when it is present in amounts below the optimum for human growth. In this case, humans cannot use the remaining amino acids to synthesize protein. If the limiting amino acid is supplied, protein synthesis can go on until another amino acid becomes limiting. Amino acids not used for protein synthesis are used as a source of calories (see Chapter 1). Another approach is to add a small quantity of a food rich in protein, such as milk powder, fish protein concentrate, soy protein concentrate, yeast preparations, and many others (Table 10.1). One can also judiciously blend one cereal product with another, such as a legume preparation, making a mixture in which the protein in one food helps balance the protein in the other. A classic example of this technique, adapted from research at the Institute for Nutrition for Central America and Panama, is illustrated in Fig. 10.3.

Table 10.1. Improvement in Nutrition Value When Lysine and Fish Flour Are Added to Wheat Flour

Product	Protein (%)	Relative Nutritional Value of Protein (%)	Utilizable Protein (%)
White flour	14	24	3
White flour + 0.2% lysine	14	38	5
White flour + 5% fish flour	17	42	7

Source: Milner (1969).

This shows the results of blending corn flour with soybean flour. The left side of the bottom line shows a mixture of 100% corn and 0% soy, the right side represents 0% corn and 100% soy, and the points in between represent blends of the two. The left-hand scale represents the rate of weight gain of rats fed mixtures of corn and soy. The right-hand scale represents the PER (protein equivalent ratio; see Chapter 11) value for the same mixtures of corn and soy. Corn meal alone promotes a low rate of weight gain and has a low PER because it is low in lysine. On the other hand, soy meal alone, although better than corn, is not optimum because it is deficient in methionine. A mixture of 40% corn and 60% soy

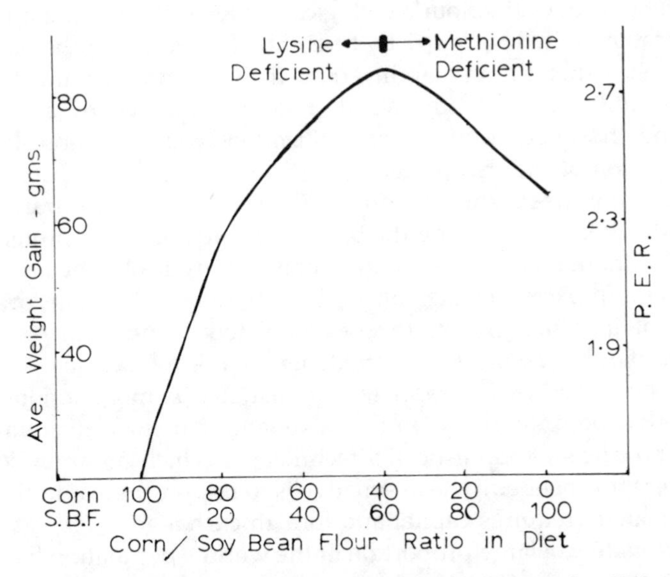

Fig. 10.3. The relationship between weight gain and PER for different mixtures of corn and soy bean flour. From Milner (1969).

provides the optimum growth rate and the highest PER obtainable with these materials. This mixture is better than either corn or soy alone because the amino acid makeup of the two supplement each other. If the amino acid makeup of any type of cereal grain and any type of legume is known, it is possible to calculate which mixtures will be superior to either one alone. If the ingredients are correct, this can be a very good way of using the existing protein supplies more efficiently. All cereals are low in lysine. Corn is also low in tryptophan. Wheat and rice are low in threonine as well as lysine. Legumes and oil-seed proteins generally, but not always, complement the cereal grains. This is probably an oversimplification; any prospective combinations should be checked by animal feeding studies. But the theory is appropriate. This concept has been used very effectively in the development of a wide variety of food formulations for infants and children in the developing countries, since it has the major advantage of utilizing foodstuffs available locally.

It is possible to improve the nutritional value of food mixtures such as the above, or for that matter of any carbohydrate source, by adding synthetic amino acids or high protein supplements. The choice will depend on cost, availability, acceptance, processing efficiency, and probably many other factors.

Another example of fortification is shown in Table 10.1. Lysine is the limiting amino acid in wheat flour, and the addition of 0.2% lysine markedly improves the flour's biological value and hence the utilization of the protein in it. This is strictly true only if there is no other source of protein in the diet, but in marginal diets the protein would undoubtedly be better utilized. The same effect could be obtained by adding fish flour, and the increase in protein utilization would be more than the mere addition of the fish protein.

The addition of synthetic amino acids or protein concentrates is undoubtedly effective in raising the biological value of cereal proteins, but it is not without some technical problems. For example, the addition of methionine in excess creates an undesirable flavor in some products. Some protein concentrates are not completely bland, and consumers may not like the flavor. These problems are not too serious, however, and can be solved by appropriate formulations. A more serious drawback is that both synthetic amino acids and protein concentrates are usually products of sophisticated technology, which may not be available where it is needed. The alternative is to import these additives, but this, of course, requires capital and foreign exchange.

Unfortunately, a large proportion of the world's population has to rely on protein from plants rather than animals, simply because of the finite supply of animal protein. The concept that protein from a source similar to the human consumer is made complex by considerations other than nutrition. Consuming protein from genetically similar sources would, of course, involve cannibalism, which is taboo by most societies. The next best source is animal protein, the use of which may be restricted by economical or sociological considerations. Plants may be a good source of protein if proper selection is made, but, on the whole, plant protein is the most dissimilar from human protein, and therefore, is nutritionally least appropriate. It is up to the ingenuity of man to seek new protein sources, to devise processes for extracting protein from various sources, and to calculate the optimum combinations of protein—in other words, to make the best possible use of the available protein for human consumption.

A tremendous amount of work has been directed towards the development of food mixtures to improve nutritional health. Most of it was sponsored by several United Nations agencies such as FAO, WHO, UNESCO and others. But it appears that nutritional improvement of infants and children is an incredibly complex situation. The efforts to improve the nutritive value of protein by enrichment of cereal grains with amino acids in the Middle East and Southwest Asia did not stimulate child growth as much as expected. Perhaps this was due to an overestimation of human protein requirements as calculated from data on laboratory rats. Infantile malnutrition may be more related to the absence of clean water than to low protein quality and to insect and pest control, sanitation and disease control than protein intake. Adequate protein and calorie intake is only a part of the larger picture.

GENETIC ENGINEERING

The phrase "genetic engineering" has captured the imagination of the consuming public, the scientific community and the financial establishment in a manner unparalleled in the past decade. The phrase means everything from conventional plant and animal breeding to DNA transfer technology. Potential benefits to mankind dwarf any other approach.

Plant Breeding

The improvement in the biological value of the protein in cereal grains may be accomplished more effectively by changing the genetic makeup of the plant. This is not usually referred to as nutrification, but the result is the same.

The corn plant in the United States is probably the outstanding example of genetic engineering, for many reasons. Not the least of these is the development of hybrid corn. Corn or maize is one of the oldest used cereals. Corn cobs have been found in caves in Mexico that date back 5,600 years. Corn pollen has been found in drill cores dated over 80,000 years old. The changeover to hybrid corn from open-pollinated varieties had profound effects on American corn production. From 1936 to 1982, the yield per acre increased 500% (from 23 to 114 bushels per acre) and the acreage decreased 25%. Corn became the major cereal crop in the United States, with three times the production and twice the dollar value of wheat. It also became a major factor in the production of red meat and poultry.

The protein content of corn and the possibilities for increasing its protein content yield some startling figures if one realizes the volume involved. For example, the 1982 corn crop in the United States was 8.4 billion bushels. At 50 pounds per bushel, this is about 210,000,000 tons. Assuming that corn is 12% protein, this represents 26,000,000 tons of protein. By comparison, world fish landings in 1982 were approximately 65 million tons. Assuming 25% protein in fish, this amounts to about 16 million tons of protein.

If one could develop varieties of corn with even 1% more protein, over 2 million tons of protein would be added. A 1% increase does not seem like much to ask from the plant breeders so a group of researchers at the University of Illinois in 1896 began to select corn varieties with a higher protein content. In 1972, after 76 generations of selections, they had increased the protein content of one variety (Illinois High protein) from 11 to 26%. But, this was an experiment in genetic diversity, not an attempt to develop commercial varieties of corn. The high protein variety was not accepted in commercial practice because the biological value of its protein was inferior to that of normal corn. Four types of protein, globulin, glutelin, albumin, and prolamine (zein) occur in corn. Zein has a low biological value because it is deficient in both lysine and tryptophane. If a corn could be produced with less zein in the kernel and more of the other three types of proteins, its biological value would be superior to that of conventional varieties.

A genetic breakthrough occurred in 1964 when Edwin T. Mertz at Purdue discovered the "opaque-2" gene. This gene was so named because the corn kernels that contained it looked opaque rather than translucent and flinty like normal kernels. It was fortunate that the kernels did look different, because this provided a simple visual marker with which to follow the subsequent development of the opaque-2 characteristics in corn varieties. It was soon discovered that the opaque-2 kernels had twice as much lysine and tryptophan as normal kernels. The nutritional potential of this finding was obvious. Research on incorporating this gene into commercial corn production was given a high priority, but commercialization was not easy. The opaque-2 gene is recessive, and yields of opaque-2 corn were 8–10% lower than the yields of conventional varieties. The gene also produced grain with a 20% higher water content. Finally, commercial production of opaque-2 started in the United States in 1969.

Another genetic breakthrough occurred in 1965, when Oliver E. Nelson at Purdue discovered another gene that also increased the proportion of lysine and tryptophan. He called this gene "floury-2." It was distinct from the opaque-2 gene because it was carried on another chromosome. Research designed to incorporate both genes into one variety has been partially successful. The lysine and tryptophan content in the new variety is higher than that in either opaque-2 or floury-2. No commercial varieties of this hybrid have been released to date, but it is only a matter of time until this is done.

The potential of high-lysine corn as an animal feed is shown in Table 10.2. The pigs fed high-lysine corn grew three-and-a-half times as fast and were twice as efficient in feed utilization. A dramatic advantage such as this is obvious only if corn is the only source of protein in the animals' diet. This is usually not the case. However, it may be more efficient to reduce the other sources of protein, such as fish meal or soybean meal, and increase the corn. The fattening of animals by grain feeding is a sophisticated computer-controlled operation accompanied by careful cost analysis in many large agribusinesses in the United States. If an overall benefit can be demonstrated, high-lysine corn will be used, in spite of its higher cost.

High-lysine corn is currently being used to feed animals in the United States. Acceptance of high-lysine corn in human nutrition is encountering much more difficulty, however. Nutritionally, it is obviously better, but in many communities corn is corn, so why pay a higher price for a

Table 10.2. Rate of Growth of Young Pigs Fed Normal and High-Lysine Corn

Feed Efficiency	Normal Corn	High-Lysine Corn
Pounds of feed/pounds gained	7/1	3.3/1
Rate of growth[a]	1	3.5
Weight gain of 35-day-old pigs after 130 days	7 lb	73 lb

[a] The rate of growth on normal corn is arbitrarily given a value of one, and other treatments are referred to this control.
Source: Harpstead (1971).

new variety? High-lysine corn will probably have to be introduced into corn-consuming cultures through government subsidies of school lunches, infant formulations, or national legislation. The recent experiments in which South American children were fed high-lysine corn are very promising. It would be a real triumph for nutrition if high-lysine corn were to replace conventional corn in the human diet.

Future Possibilities for Genetic Improvement

The discovery of genes that improved the protein quality in corn for human nutrition set off a research race to do the same thing for other crops. In 1973, scientists at Purdue announced the discovery of a gene in sorghum that increased the total protein content in that plant by 30–40% and doubled the lysine content. Sorghum is the "poor man's corn," in that it will grow on land that is too dry for corn. Sorghum is the principal cereal for over 300 million people, and the new discovery is of great potential benefit to them. There are indications that a high-lysine gene exists in barley, but the prospects are less promising for the other major cereal crops such as wheat, rice, rye, and millet. Wheat and rice have been the subject of intense genetic research, but this research has been concerned primarily with yield and resistance to disease. The other grains have lagged behind. The major carbohydrate crops, such as cassava in the tropics, have been virtually untouched by this type of research. The benefits of genetically improving such crops are great, but cassava, for example, is such a poor protein source that it may be more efficient to grow it only as a source of carbohydrates. Perhaps we should explore both avenues.

One may well ask why the emphasis in genetic improvement is on protein. The answer is probably that protein has been judged to be the most important nutrient. Cereal grains are already excellent sources of carbohydrate, so the genetic research there has been to change the type of carbohydrate. One gene in corn called "waxy" produces a corn with a high content of amylopectin. Amylopectin is a type of starch with a highly branched structure, which makes the starch more easily digested or fermented. Another gene in corn produces amylose, a starch with a long straight-chain structure. Amylose is ideal for spinning into fibers or forming into cellophane-like sheets. It can be made water soluble, or insoluble, permeable to oxygen, and even with great shock resistance. The film that is made from it is superior to many films on the market. And it is edible: one could eat an amylose package as well as the food in it if one wanted to overlook the sanitary problems.

Corn has also been subjected to genetic research on oil type and content. For example, the Illinois experiment with protein on the original Burr white variety was also used for oil research. After 76 generations of selection, the oil content had been raised from 4.7 to 18%. This experimental variety called Illinois High Oil was never accepted commercially because as the oil content went up, the overall yield went down.

The composition of the oil in corn seeds would seem to be very important because of the association of saturated fats with heart disease. Apparently, it was purely a matter of chance that corn with a high content of the desirable linoleic and oleic fatty acids and less of the less desirable linolenic acid was developed. There has been little research on changing the chemical composition of the oil by genetic means, probably because there has been little reason to do so. The present makeup is considered good—that is, high in "polyunsaturates." If a more solid fat was desired, it could easily be produced from corn oil by the chemical process of hydrogenation.

Changes in the content of fatty acids in oil have already been accomplished in rapeseed oil in Canada. One of the fatty acids, erucic acid, is considered undesirable because it causes heart muscle lesions in rats, but apparently not in humans. Consequently, plant breeders developed varieties with a low content of erucic acid and the oil from them is called "canbra oil" or sometimes LEAR (low erucic acid rapeseed). There are also undesirable compounds called glucosinolates in rapeseed oil. Varieties which have a low content of both erucic acid and glucosinolates are used to produce "canola oil."

There is plenty of room for genetic research designed to improve the nutritive value and other characteristics of many of the world's food crops, but such research is slow and difficult. The scientists who discovered high-lysine sorghum estimate that it may take 15 years to develop appropriate commerical varieties. Genetic research may represent many man-years of sophisticated effort, often for meager returns. But the stakes are high, and once achieved, the results—except, of course, for disease resistance—are usually permanent. There is also much room for genetic selection. Man uses very few of the 250,000 known species of higher plants. There has been only one new cereal grain introduced in recent history. This man-made grain, triticale, is a cross between wheat and rye. Triticale produces higher yields than either wheat or rye, and it also has a high protein content. Research on producing an acceptable bread from triticale has been slow, but the pace has increased in the past 10 years. Undoubtedly, triticale will take its place among the major cereal grains of the world.

Other Forms of Genetic Engineering

The use of conventional plant and animal breeding and solution of desired cultivars and species has been responsible for the major developments in both plants and animals. Yet scientists are always searching for other ways to create new and different plants and animals. Mutations can be created in a number of ways. For example, treatment of growing tissue with the chemical, colchicine, has been used to double the number of chromosomes in a plant. Such diploid plants are usually bigger, possibly more vigorous, but certainly different from the original plant. The rutabaga, sometimes called yellow turnip, is a good example of a diploid vegetable.

Another way to create mutations is by irradiation. After World War II and the development of nuclear reactors, it was possible to irradiate plants on a large scale. It was anticipated that some of the mutants would show beneficial properties. This program was disappointing, however; no major benefits did develop.

A third approach has been cell fusion. With this method, the outer layer of a growing cell is stripped away and a cell of dissimilar species encouraged to fuse together. This was the way the much publicized "sunbean" plant was created. In this case, cells from a sunflower were fused with those from a French bean in the hopes that the resulting plant would have the vigor of a sunflower and the protein synthesizing ability of the bean. Actually, the sunbean is still living as a piece of tissue but it is not recognizable as a plant. Another product of this technology is the "pomato." The idea of creating this hybrid was to produce potatoes beneath the ground and tomatoes above. Actually, the pomato may not be too far-fetched because the two plants are closely related species. In reality, a true pomato does exist but, to date, it has produced neither potatoes nor tomatoes. Pomato plants offered for sale are usually a fraud. What one gets is either two plants growing together or, a tomato plant grafted onto a potato plant.

Cell fusion experiments have some fascinating possibilities for animals. For example, researchers in Cambridge, England have produced eight geeps. A geep is a cross between a goat and a sheep. A fertilized cell from a goat and a sheep were fused and planted in the womb of a surrogate mother. The cells in contact with the placenta of the mother were those of the same species as the mother, so the mother's immune system did not recognize the cells as being foreign. Each geep had four parents and showed recognizable traits of both goats and sheep. The researchers say they are not trying to develop a "sweater-and-cheese" producing farm animal, but this technology may open up many new possibilities for crosses between species.

The form of genetic engineering that has really caught the imagination of both the scientists and the financial community is the "recombinant DNA transfer" approach. The amazing progress in basic genetic studies has allowed scientists to locate the genes in the chromosomes that are responsible for a myriad of reactions in plants and animals. After locating the appropriate gene, it can be removed and relocated in a chromosome of another organism. This organism will then proceed to do whatever the gene was programmed to do in the original cell. For example, the gene that controls the production of insulin in the pancreas of rats was located. After isolation, the chemical structure of the gene was determined and the gene was synthesized using a commercial "gene machine." Actually, the researchers synthesized the gene structure to make human insulin rather than rat insulin, which differ in one chemical grouping. The synthetic genes were introduced into a microorganism which was persuaded to grow and produce insulin. The Eli Lilly Co. has built the first commercial plant to make insulin by recombinant DNA methods, and insulin is now available. Insulin is the hormone that controls the utilization of carbohydrates in normal animals and is deficient or absent in diabetics. Human diabetics must inject insulin daily to make up for this insufficient supply. Most insulin-dependent diabetics use insulin that has come from animal pancreas and there has been some question about potential supply. The DNA approach could certainly remove this uncertainty.

Research in other areas of diabetes is moving fast. Two other developments may obsolete the DNA process even before it gets well established. One group of workers has found a way to transplant insulin-producing cells from a normal animal into a diabetic animal. Previously, transplants were destroyed immediately by the animal's immune system. One of the signals that triggers the immune response is provided by cells called passenger leukocytes. The researchers removed the leukocyte cells from the transplants and the tissue was not rejected. Another research team has placed insulin cells within small spheres and implanted the spheres in the host. The pores in the spheres allow the insulin to diffuse out yet are too small to admit the host's defensive cells. The spheres themselves were not rejected because the host does not recognize the carbohydrate coating as being foreign. Research on diabetes appears to be on the brink of exciting breakthroughs and even more exciting, it appears that the same approach can be used for a number of other disases.

The recombinant DNA approach has been used to make interferon, human growth hormone, a vaccine for foot and mouth disease in cattle, rennin for cheese production, several enzymes, flavor compounds, and specialty chemicals. The potential appears unlimited. Perhaps the ultimate challenge for the plant scientists is to create a corn plant that can fix its own nitrogen from the atmosphere. This involves transferring a packet of genes from a legume to a corn plant. Apparently this process is incredibly complicated. By comparison, insulin production is controlled by one gene, whereas nitrogen fixation involves at least 17. They have to be transferred along with the genetic control segments, which turn each gene on and off in the proper sequence. Insulin production, as impressive as it may be, when compared to nitrogen fixation has been likened to Henry Ford's Model T and the spaceship Discovery. Yet the stakes are so high that this is a very active field of research.

BIBLIOGRAPHY

High-Quality Protein Maize. New York: John Wiley & Sons (Halstead Press, 1975).

HARPSTEAD, D. D. 1971. High-lysine corn. Sci. Am. *225,* 38.

INGLETT, G. E. 1970. Corn: Culture, Processing, Products. AVI Publishing Co., Westport, Connecticut.

MILNER, M. 1969. Protein Enriched Foods for World Needs, p. 42. American Association of Cereal Chemists, St. Paul, Minnesota.

RECHEIGL, M. 1983. Handbook of Nutritional Supplements, Vol. 1, Human Use. CRC Press, Cleveland, Ohio.

SCRIMSHAW, N. S., and ALTSCHUL, A. M. 1971. Amino Acid Fortification of Protein Foods. MIT Press, Cambridge, Massachusetts.

11

Nutritional Labeling Impact on Nutrification

The food nutrification program received a real stimulus in 1973 with the passage of the nutritional labeling laws in the United States. This was an indirect effect, however, since these laws were passed for other reasons.

The nutritional labeling legislation is an outgrowth of efforts to make information on nutrition and food composition more available to the consumer. This in itself is a worthy aim, since it really was difficult for the average consumer to judge just what nutrients a food does contain. It was even more difficult to interpret the significance of some of these nutrients. This situation was not helped by the wealth of misinformation being promoted by popular periodicals and books. Some of these have a very wide circulation and cater to popular interest by being sensational without having to be accurate. Because of this, it was very difficult for the consumer to judge food content and nutritive value. The nutritional labeling laws were designed to help with at least part of this dilemma—the determination of nutrients in a food.

The nutritional labeling laws in the United States became effective on July 1, 1975, and they are very specific as to the information that must be stated on the label:

1. The serving size and the number of servings per container.
2. The calorie content to the nearest ten calories per serving.
3. The protein content to the nearest gram per serving, and also as a percentage of the RDA (Recommended Daily Allowance; see pages 11–13.
4. The carbohydrate content to the nearest gram per serving.
5. The fat content to the nearest gram per serving.

The serving size is determined by agreement between the manufacturer and the FDA authorities. Some examples are obvious—for instance, one muffin—but others such as catsup, are less so. The concept of serving size has generated considerable discussion but no real problems. Calorie and carbohydrate content are determined by standard analytical methods.

PROTEIN CONTENT

Evaluating the protein content was a real problem, in view of the difficulties in describing protein quality. The final compromise was to allow an RDA of 45 for protein if the PER of the protein is equal to or greater than that of casein, a high-quality milk protein. If the protein quality of the food is lower than that of casein, the RDA is 65 grams. If the PER of the protein is 20% or less of the PER of casein, no label declaration could be made. As discussed previously the PER (protein equivalent ratio) is determined by feeding food containing the protein in question to a group of young growing rats. The weight gain (in grams) and the protein fed (in grams) are determined, and the ratio—weight gain divided by grams of protein—is the PER. A similar group of rats is fed casein and a PER is determined. This value for casein is arbitrarily set at 2.8, and the PER for the protein underquestion is adjusted accordingly. The PER is the simplest but not necesarily the best method of establishing protein quality. The overall solution to the problem of labeling protein as a percentage of the RDA is a compromise of many complex considerations, but it is probably one of the best compromises available.

FAT CONTENT

Fat content was the subject of considerable discussion, in view of the importance of the quantity and type of fat in diseases involving the circulatory system. If a manufacturer wishes to make a nutritional claim for fat—i.e., concerning saturated and unsaturated fat—the amounts of saturated and unsaturated fatty acids must be stated in grams per serving. These two statements cannot be used if the food contains less than 10% fat or 2 grams per serving. Obviously, if nutritional claims for fat are to be made, the food should have a reasonable fat content. The cholesterol content may also be stated in grams per serving. If either fatty acids or cholesterol, or both, are stated, the label must have the following disclaimer: "Information on fat (and/or cholesterol, where appropriate) is provided for individuals who, on the advice of a physician, are modifying their total dietary intake of fat (and/or cholesterol, where appropriate)." This disclaimer may have about the same degree of success as the medical disclaimer on packages of cigarettes.

VITAMINS AND MINERALS

The listing of fat, protein, and carbohydrate in terms of grams per serving is easy to understand, but the method of stating vitamin and mineral content was more difficult. Several methods were examined in great detail and with some emotion. The decision was to use the United States Recommended Daily Allowance (US RDA). These units were based on the National Academy of Sciences-National Research Council Recommended Dietary Allowances (NAS/NRC RDA), which were first determined in 1941 and have been revised eight times. The latest revision appeared in 1980 (see Table 1.2). The NAS/NRC RDA values represent the best nutritional information available on the daily intake of vitamins and minerals needed to maintain good health. The NAS RDA replaced the older and obsolete Minimum Daily Requirements (MDR), which were based on the minimum intake necessary to prevent deficiencies and related illnesses. Individuals vary in their need for each nutrient, so the NAS RDA requirement was set at two standard deviations above the average requirement. This means that 97.5% of the population are included in the NAS RDA requirement. The NAS/NRC RDA recommendations were broken down into a number of categories, such as age, sex, lactation, and pregnancy, and obviously involved too much detail to be declared in the limited space available on the label. Consequently, the US RDA values (see Table 11.1) represent a selection (usually the highest of the NAS/NRC RDAs). There are, however, other nutrients. Table 11.2 lists nutrients for which an estimated safe and adequate intake has been established. These estimates are based on a smaller data base than the NAS RDA figures. The choice of US RDA units for label declaration is a good one, even though the reader cannot determine the actual content of vitamins and minerals from the label. It is easy to calculate this information by referring to Table 11.1. The US RDAs are very liberal requirements.

FDA regulations require that the content of seven vitamins and minerals be stated. These are the vitamins A, C, thiamin, riboflavin, and niacin, and the minerals iron and calcium. A manufacturer cannot choose which nutrients he will put on the label. If he puts any, he has to put the seven listed above. Nine others may be declared if the manufac-

Table 11.1. The US RDA Values for Human Nutrition

Nutrient	Units
Vitamin A	1000 μg RE[a]
Vitamin C (ascorbic acid)	60 mg
Thiamin (Vitamin B_1)	1.5 mg
Riboflavin (Vitamin B_2)	1.7 mg
Niacin (Vitamin B_3)	19 mg
Calcium	1.2 g
Iron	18 mg
Vitamin D (cholecalciferol)	10 μg cholecalciferol
Vitamin E (tocopherol)	10 mg (TE)[a]
Vitamin B_6 (pyridoxine)	2.2 mg
Folic acid (folacin)	0.4 mg
Vitamin B_{12} (cobalamin)	3 μg
Phosphorus	1.2 g
Iodine	150 μg
Magnesium	300 mg
Zinc	15 mg

[a] RE = retinol equivalents, TE = tocopherol equivalents. See pp. 72 and 76.

turer so wishes. These are vitamin D, vitamin E, folacin, vitamin B_6, vitamin B_{12}, zinc, magnesium, phosphorus, and iodine. All 16 shall be stated as the percentage of the US RDA that is contained in one serving. No other vitamins or minerals can be claimed on the label. This limitation is to prevent manufacturers from claiming nutrients that the scientific community believes to be unnecessary in the human diet. The sodium content in milligrams per serving may also be stated on the label.

Table 11.2. Estimated Safe and Adequate Daily Dietary Intakes for Adults of Additional Selected Vitamins and Minerals

Nutrient	Units
Vitamin A	70–140 μg
Biotin	100–200 μg
Pantothenic acid	4–7 mg
Copper	2–3 mg
Manganese	2.5–5.0 mg
Fluoride	1.5–4.0 mg
Chromium	50–200 μg
Selenium	50–200 μg
Molybdenun	15–500 μg
Sodium	1.1–3.3 g
Potassium	1.9–5.6 g
Chloride	1.7–5.1 g

FOODS, SUPPLEMENTS, AND DRUGS

The labeling laws as originally presented were very specific about the classification of products. Foods with added nutrients that represent up to 50% of the US RDA were considered ordinary foods. Products with added nutrients that represent between 50 and 150% of the RDA were considered "dietary supplements" and had to be labeled as such. Products with nutrients over 150% of the RDA were considered "drugs" and must be so labeled. Because the regulations governing the sale of drugs are more stringent than those for food, it is unlikely that many food manufacturers would choose to label their foods as drugs. Interestingly, several of the breakfast cereals that contain 100% of the RDA of a number of nutrients are now labeled as dietary supplements. One exception to the drug ruling is that if the nutrient occurs naturally in the food at levels above 150% of the RDA, the food need not be labeled as a drug. The FDA has recently relaxed the ruling that preparations with over 150% of the RDA be called drugs. They can now be marketed as dietary supplements.

THE VOLUNTARY ASPECT

Nutritional labeling laws are voluntary, with two exceptions. If a manufacturer adds nutrients or makes nutritional claims, the labels must state the amounts of the five vitamins and two minerals. If a manufacturer does not add nutrients and does not make nutrient claims, then he does not have to label the product in this manner. The concept of voluntary labeling makes good sense because it gives the manufacturer the choice whether to label or not to label some food products. This may soften the economic impact for complete, mandatory nutritional labeling, which may come later. Proponents of the present legislation are betting that pressure from competition in the marketplace will persuade most manufacturers to comply with the new labeling laws even if they do not add nutrients.

Nutritional labeling laws will have an indirect effect on nutrient content. Marketing departments will be hard pressed to promote a product with an asterisk on the label that denotes "contains less than 2 percent of the US RDA of these nutrients" and will bring pressure on the producers to make their products appear better nutritionally. This has been referred to as "the horsepower race," and it might well be that. However, current legislation is sufficient to curb abuses in this area. Modest additions of nutrients would probably be an advantage from a marketing point of view and marketing managers may encourage this development.

Purveyors of organic and health foods may be hurt by the nutritional labeling laws. They sell food to the public and therefore should comply with the food laws. They usually make nutritional claims and thus will have to label the nutrient content of their products. For the first time, the consumer will have the opportunity to compare the nutrient content of health foods, natural foods, and commercial foods. Some of the comparisons are very interesting for example, some of the nonfortified natural breakfast cereals as compared to the fortified cereals.

The nutritional claims for some foods will be modified by the new laws, since any nutritional claim must be substantiated by good nutrition data. For example, the following claims will not be allowed.

1. That the food, because of the presence or absence of certain dietary properties, is adequate or effective in the prevention, cure, mitigation or treatment of any disease or symptom.
2. That a balanced diet of ordinary foods cannot supply adequate amounts of nutrients.
3. That the lack of optimum nutritive quality of a food by reason of the soil on which the food was grown, is or may be responsible for an inadequacy or deficiency in the quality of the daily diet.
4. That the storage, transportation, processing or cooking of a food is or may be responsible for an inadequacy or deficiency in the quality of the daily diet.
5. That the food has dietary properties when such properties are of no significant value or need in human nutrition.

The elimination of claims such as these will help considerably to prevent the spread of misinformation on human nutrition. At the same time, it will not prevent the use of legitimate nutrition claims.

Some people have suggested that the amounts of nutrients on a label be obtained from tables of nutritional data. This idea was not adopted in 1975 because the existing tables were not adequate for the purpose. Since then, many groups, and particularly the U.S. Department of Agriculture, have mounted a very large campaign to make nutrient data banks much more complete. It should be possible in the future to make reasonable estimates of nutrient content of a product for nutritional labeling purposes from these nutrient data banks. Others have objected to the inclusion of only seven mandatory nutrients of the 33 or so that are known to be required for optimum human nutrition. The answer has been that if the seven are supplied, there is every likelihood that the other 33 will be present also. This is a rather tenuous assumption, particularly if most of the seven are supplied by fortification. The answer to this question is continued surveillance of national nutrient intake to see whether some groups of consumers are actually at risk from inadequate intake of one or more nutrients. If this is found to be the case, it could be remedied by encouraging changes in food supply and consumption, or even by legislation.

THE OUTLOOK

The nutritional labeling laws are a very complex package of regulations. They deal with an exceedingly wide array of food products and are open to many interpretations, most of which are beyond the scope of this chapter. In the area of enforcement, for example, an FDA spokesman was asked how one should obtain the data for the content of nutrients in a food. He answered facetiously, "Use a table of random numbers—but be sure you are correct." He really meant that the methods of analysis are up to the manufacturer, but the results should agree with the official methods accepted by the FDA and the USDA.

The implementation of the nutritional labeling laws unquestionably adds to the cost of food to the consumer. There is a large cost to the manufacturer for the initial data, and a much smaller continuing cost entailed by making sure the data are current. It is hoped the overall cost to the consumer per unit of food is very small. It will take a tremendous educational campaign to enable consumers to take optimum advantage of the labeling laws. The cost of compliance and the cost of education will have to be balanced against the advantages to consumers. Yet the expenditures are inevitable because more processed foods are being included in the U.S. food supply. For example, in 1941 an estimated 10% of our food was highly processed. The figure was about 50% in 1975, and about 55% in 1984.

There has been recent considerable discussion about changing the manner in which nutritional labeling is done, for example, using bar or pie charts, or adjectives such as "good source," "fair source," etc., instead of percentages. Many other countries use these alternative formats. Consumer surveys have shown that consumers pay little attention to the thiamin, riboflavin, and niacin requirements. Perhaps they should be dropped from the label. Hopefully, there will be no major changes in the nutritional labeling laws until sufficient time has elapsed to judge the successes or failures of the present law. One important change has already happened. In 1985, sodium content will be mandatory for those using nutritional labeling. In addition, for those not using the nutritional labeling format, sodium content can be declared without triggering the full nutritional labeling format. This law is in response to the public concern for sodium as related to high blood pressure.

Against this background, the old concept of the "basic four" food groups (dairy products, fruits and vegetables, cereal products, and meats) has become inadequate for nutritional guidance. We will have to superimpose more sophisticated yardsticks on top of the basic four. One step in this direction is the declaration of nutrient content on the labels of foods. If the educational goal is beyond the capacity of the American public, then the food fortification program should progress to the point where it would be very difficult to maintain an unbalanced diet. Perhaps we should proceed along this route in any event. Federal legislation is moving toward national nutritional-quality guidelines, and the net effect is to encourage nutrification. There will always be individuals who prefer to eat only one or two foods. One cannot legislate against this belief, but one can educate.

BIBLIOGRAPHY

DEUTSCH, R. M. 1975. Nutrition Labeling: How It Can Work for You. National Nutrition Consortium, Bethesda, Maryland.

Nutrition Labeling: Tools for Its Use. Agriculture Information Bulletin 382. U.S. Department of Agriculture, Washington, DC.

12

Dietary Goals

Americans are considered to have one of the most varied, abundant, safest and appealing food supply systems in the world. Also in this list, hopefully, is a very nutritious food supply, combined with a national awareness of the importance of nutrition. Associated with the food supply and nutrient delivery system are excellent public health programs. But, the concept of unlimited top quality medical care for everyone is a matter of economic concern. In 1970, the cost of health care in the United States for 200 million people was about $75 billion. In 1982, for about 230 million people, the bill was $322 billion. This is an increase of 430% in 12 years. Put another way, health care was 5% of the gross national product (GNP) in 1960, 7½% in 1970 and 10% in 1982. In 1982, health care cost $1365 per person. Even allowing for inflation, the rate of increase in health care was far outstripping increases in the GNP, and obviously this situation could not continue indefinitely. This was the political background behind the recent emphasis on Dietary Goals. What could possibly be more appealing than the prospect of lowering the costs of national health care by improvements in the diet. Healthy people need less medical care.

NUTRITIONAL EDUCATION

The Dietary Goals are another way of encouraging consumer awareness of good nutrition. This concept has taken several forms and has been around for a long time. In 1944, food was divided into 13 groups and consumers were encouraged to partake of each group to get a varied diet. The 13 groups were then reduced to 9, then to 7, and later to 4. In 1982, 5 groups were recognized. They are called the "Basic Five" or, sometimes the "four plus one."

1. Milk and milk products
2. Fruits and vegetables
3. Meat, poultry, fish, and eggs
4. Bread, flour, and cereals
5. Fats and sweets

The last group was added to conform to the reality of what people were actually eating. This placed the USDA in the peculiar position of appearing to recommend the consumption of fats, candies, soft drinks, etc. This dilemma was resolved by promoting the fifth group only as a source of needed calories, whereas actually, recommendations are usually to reduce the consumption of fats and sweets.

The "Basic Five" concept is a simple and effective way to promote nutritional knowledge in schools and consumer groups. It served very well in past decades when food staples were more recognizable. However, as we progress further down the road towards engineered, or highly processed foods, or even some ethnic foods, the classification becomes rather fuzzy. For example, how would you classify a taco? The vegetable portion would put it into Group 2, whereas the meat portion would be in Group 3, and the shell would be in Group 4. Effective as the Basic Five may be as a teaching tool, we need more. This is the reason behind the nutritional labeling program described in Chapter 11. The nutritional labeling program listed nutrients, but we eat food not nutrients. Therefore, both approaches are needed.

The preceding chapters were devoted to food supply, improvement of nutrient content of food, and programs to enable consumers to judge nutrient control. Yet, the success of all of these efforts depends on the ability of scientists to judge what is an optimum diet and to convey this information to consumers. The challenge is well understood and the political and economic motivation is unparalled. Yet in this very controversial area, the delivery leaves something to be desired.

The goal of an optimum diet is the same now as at the dawn of civilization. If one asked our colonial forefathers what they expected from food, the answer would probably be "enough food to prevent starvation." Fifty years ago, the answer might have been, "I know I will get enough food, I want to be assured that I will get enough of those newly discovered vitamins." Today, the answer would probably be "I know I will get enough vitamins, minerals and calories (probably too many calories). I want a diet that will enable me to live well and enjoy the good life." Articles in the popular press describe diets to cure cancer, diabetes, arthritis, heart disease, stroke, and intestinal problems. Diets are available to make one smarter, able to talk better, run faster, look better, enjoy sex, and almost anything you wish. Many of these claims are frivolous, but the claims for medical benefits are serious. Some of them may be possible, but the professionals are going to have difficulty delivering on many of these topics.

DIETARY GOALS

The relationship between diet and disease has been studied in a number of official reports. The one that perhaps has had the most publicity was entitled "Dietary Goals for the United States." It was published in 1977 by the Select Committee on Nutrition and Human Needs headed by Senator George McGovern. A number of eminent scientists were asked to give their views on diet, nutrition, and health to the Senate Committee. The 83-page report includes fascinating data on historical food consumption in the United States, and makes a number of recommendations for optimum health (Fig. 12.1). One major recommendation was to reduce the fat content of the diet from 42% of calories to 30%. Another was to increase the carbohydrate content from 46% (22% complex carbohydrates, 6% natural sugars, 18% refined and processed sugars) to 58%, of which only 10% was to be refined and processed sugars. This report created a real controversy, with ardent supporters on both sides. The committee was disbanded in 1977, and responsibility for nutritional concerns was given to two government departments, the Department of Agriculture (USDA) and the Dept of Health and Human Services (USDHHS). These two departments produced in 1979 another report entitled, "Nutrition and Your Health. Dietary Guidelines for Americans." It was similar to the previous report, but was less controversial. The recommendations can be summarized as follows:

1. Eat a variety of foods daily
2. Maintain ideal weight
3. Avoid too much fat, saturated fat, and cholesterol
4. Eat foods with adequate starch and fiber
5. Avoid too much sugar
6. Avoid too much sodium
7. If you drink alcohol, do so in moderation.

Another report also appeared in 1979, from the U.S. Surgeon General's office, entitled "Healthy People." It supported the low-fat, high carbohydrate theme of the earlier report. In 1979, the National Cancer Institute, previously uncommitted on dietary policy, in a "Statement on

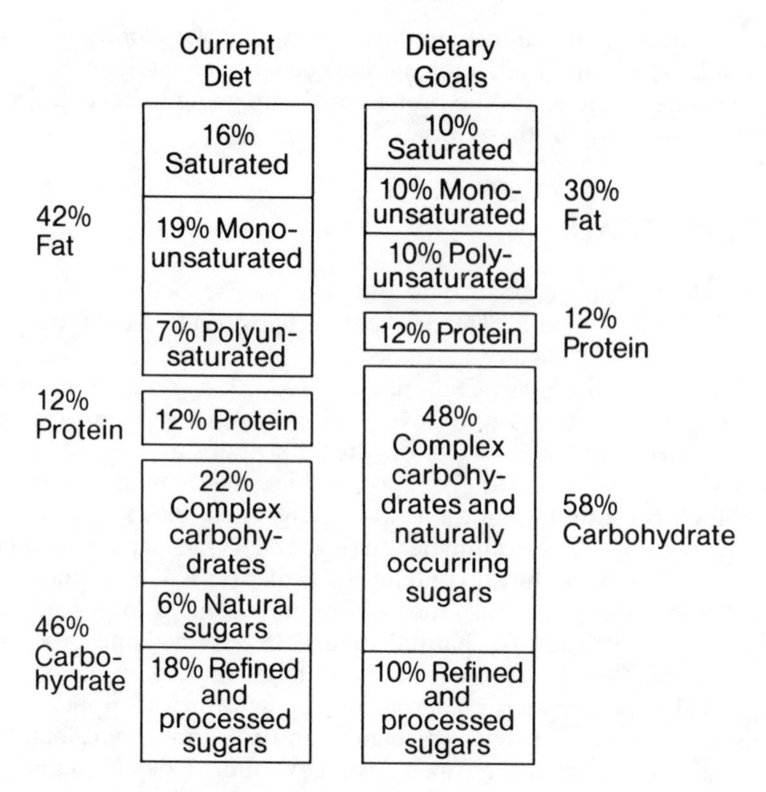

Fig. 12.1. Recommendations from the "Dietary Goals for the United States".

Diet, Nutrition and Cancer," also supported the low-fat recommendations.

Two reports probably received the most public attention. One was entitled "Toward Healthful Diets" published in 1980 by the National Academy of Sciences. Its recommendations were as follows:

1. Select a nutritionally adequate diet from the foods available, by consuming each day appropriate servings of dairy products, meats or legumes, vegetables and fruits, and cereal and breads.
2. Select as wide a variety of foods in each of the major food groups as is practicable in order to ensure a high probability of consuming adequate quantities of all essential nutrients.
3. Adjust dietary energy intake and energy expenditure so as to maintain appropriate weight for height; if overweight, achieve

appropriate weight reduction by decreasing total food and fat intake and by increasing physical activity.

4. If the requirement for energy is low (e.g., reducing diet), reduce consumption of foods such as alcohol, sugars, fats, and oils, which provide calories but few other essential nutrients.
5. Use salt in moderation; adequate but safe intakes are considered to range between 3 and 8 g of sodium chloride daily.

The above recommendations are pretty general since the committee members apparently did not feel that there was sufficient evidence linking diet with diseases to warrant sweeping changes in the american diet. The second report, also from the National Academy of Sciences in 1982 was entitled "Diet, Nutrition and Cancer." Its members did believe that there was sufficient evidence to warrant their conclusions as follows:

1. *Eat less fat.* Fat consumption should be reduced 25%. The committee found insufficient evidence to recommend changes in the type of fat ingested (e.g., saturated vs. polyunsaturated).
2. *Consume adequate amounts of vitamins A, C, and selenium.* Vitamin supplements are discouraged because excessive vitamin A and selenium are toxic. The best sources of these nutrients are fruits, vegetables, and whole grains.
3. *Eat less cured meat products and smoked products.* Such products are of concern because of the possible formation of nitrosamines and complex cyclic hydrocarbons.
4. *Drink alcohol only in moderation.* Smokers, especially, were advised to restrict alcohol consumption.

The NRC recommendations are very similar to the USDA nutritional guidelines and to the recommendations of the "McGovern Report."

The National Academy of Sciences found itself in the position of having two committees looking at the same data and coming to different conclusions.

The latest round in the diet and disease controversy was a set of guidelines, published in 1984, by the American Cancer Society. Their recommendations are as follows:

1. Avoid obesity
2. Cut down on total fat intake
3. Eat more high-fiber foods such as fruits, vegetables, and whole-grain cereals

4. Include food rich in vitamins A and C (fruits and vegetables) in the daily diet
5. Include cruciferous vegetables such as cabbage, broccoli, brussels sprouts, kohlrabi and, cauliflower in the diet
6. Be moderate in consumption of alcoholic beverages
7. Be moderate in consumption of salt-cured, smoked and nitrite-cured foods.

There is a theme that occurs in nearly all of the recommendations listed above. It is a reduction in fat, an increase in fiber, and an increase in complex carbohydrates. The decrease in fat content of the diet was not taken very kindly by the beef, pork, egg, chicken, and dairy industries because conforming to the guidelines, meant a decrease in consumption of their products. It did seem rather strange to some consumers, who had been taught for generations that meat, eggs, and dairy products were highly desirable, now to be told that they were not so good after all. The changes in eating habits needed to achieve the USDA Dietary Guidelines for Americans, are listed in Table 12.1.

Diets which would meet the Senate Dietary Goals were calculated by Betty Peterkin of the USDA. A typical example is as follows:

2½–3 cups of cereal
13 slices of bread or other equivalent baking products
2½ cups vegetables or fruit
½ egg
5 oz lean meat, poultry, or fish
1½ cups skim milk

The above diet is not exactly primed to whet the appetite. It may well be that the goals went too far too fast, and the public was not ready to accept such radical changes. Perhaps a number of consumers will move toward the goals even to a small degree. The consensus of scientific opinion in the U.S. seems to be that this would be a good idea.

Table 12.1. Means to Achieve the USDA Dietary Goals

Increase	Decrease
Fruits, vegetables, whole grains	Sugar
Poultry, fish	Red meat
Nonfat milk	Whole milk
Polyunsaturated fat	Total fat
	Butterfat, eggs
	Salt

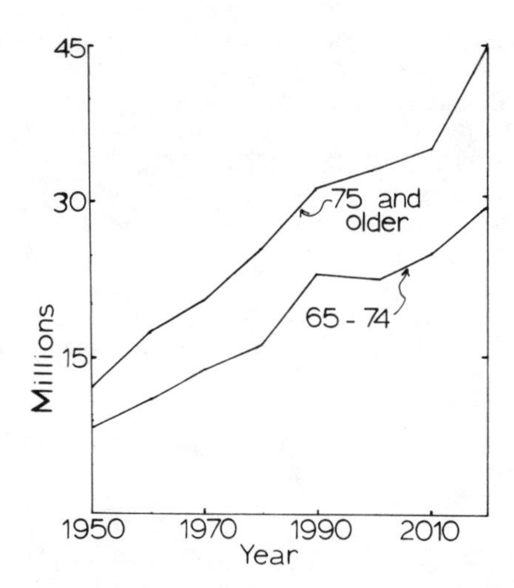

Fig. 12.2. The predicted increase in the number of elderly people in the United States.

CONCLUSIONS

Many Americans are concerned about the relationship between diet and disease or, in a more positive sense, between diet and good health. Perhaps four reasons are paramount. One is that the number of Americans living longer is increasing (Fig. 12.2). Second, the economic costs of health care are increasing at an alarming rate. Third, the population is more sedentary and requires fewer calories than in past generations. This requires more care in choosing diets in order to get optimum nutrition with fewer calories. Fourth Americans expect a better quality of life.

The relationships between diet and health are real, but at the present time very controversial. Many of the dietary recommendations for good health are based on data from animals and epidemiological data from humans. Evidence for the relationship between diet and disease is contradictory, as discussed in Chapters 1 and 5. Yet, we anticipate that a consensus will develop in many of the current controversies in the next decade.

BIBLIOGRAPHY

ANON. 1979. Healthy People: The Surgeon General's Report on Health Promotion and Disease Prevention. DHEW (PHS) Publ. No 79-55071, Department of Health, Education and Welfare, Washington, DC.

ANON. 1980. Toward Healthful Diets. Food and Nutrition Board, National Research Council, National Academy of Sciences, Washington, DC.

COMMITTEE ON DIET, NUTRITION AND CANCER. 1982. Assembly of Life Sciences, National Research Council, National Academy of Sciences, Washington, DC.

DIETARY GOALS FOR THE UNITED STATES. 1977. Select Committee on Nutrition and Human Needs, 2nd Edition, U.S. Govt Printing Office. Washington, DC.

NUTRITION AND YOUR HEALTH. 1979. Dietary Guidelines for Americans. U.S. Departments of Agriculture and Health and Human Services. U.S. Govt Printing Office, Washington, DC.

UPTON, A. C. 1979. Statement on Diet, Nutrition and Cancer. National Cancer Institute. Hearings of the Senate Subcommittee on Nutrition. Washington, DC.

13

Food Safety

The question, "Is the food safe?", usually evokes an emotional response. Everybody has a right to a safe and wholesome food supply. Yet the question of food safety is a complicated one.

CHEMICALS IN THE FOOD SUPPLY

The concept of food safety has many facets. Perhaps the most obvious, and certainly the most prevalent problem of food safety is microbiological food poisoning (see Chapter 4). This is a serious problem in nearly every country, but another aspect of food safety is primarily concerned with chemicals. Some chemicals should be there and some should not. Chemicals in our food supply come from six sources. These are industrial accidents, intentional fraud, natural compounds, food additives, pesticides, and environmental pollutants. All require a different concept of food safety and a different approach.

Industrial Accidents

The problem of industrial accidents is difficult and complex. In our technologial society there will always be accidents. We have to gear our laws and responses so that these unfortunate occurrences are minimized. One of the most unfortunate industrial accidents involving food occurred in May 1973. The Michigan Chemical Co. accidentally shipped several 100 lb bags of Firemaster, a brand of fire-retardent chemicals known as polybrominated biphenyls (PBBs), to a Farm Bureau Services feed mill. The firm had intended to ship a mineral nutrient preparation (magnesium oxide) but sent the PBB bags by mistake. The feed mill incorporated the PBB into conventional feed mixes and the chemicals were widely distributed in Michigan. Problems began to surface in September when dairy herds had decreased milk production. In October, the cattle showed obvious signs of sickness, but PBB was not identified as the cause until April 1974. By that time the PBBs were widely distributed in the environment. It was the worst single case of livestock poisoning in American agriculture. More than 30,000 cattle, 1,500,000 chickens, 4,600,000 eggs, 4,000 hogs, 750 tons of feed, thousands of pounds of cheese, ducks, pheasants, etc., were condemned. The animals had PBB contents above the action levels[1] set by the FDA. Fortunately, monetary compensation was available. However, millions more had levels below the FDA action level, and nothing could be done. In addition, thousands of humans were contaminated with PBBs, which caused much mental anguish. Fortunately, to date, follow-up surveys for human problems have been negative.

Intentional Fraud

The situation where fraud is intentional is more straightforward. This requires the passage and enforcement of adequate laws to punish unscrupulous operators, particularly in cases involving human health. Perhaps the worst recent case known of intentional fraud occurred as the "cooking oil" incident in Spain. In 1981, something caused the death of 259 people and made over 20,000 seriously ill. Everything was blamed, including strawberries, green vegetables, asparagus, birds, dogs, cats, germ warfare, etc., but the cause was finally traced to cooking oil. An enterprising firm had imported denatured rapeseed oil intended for industrial use and reprocessed it for human use. After adding soy oil and pork fat, the concoction was sold at a 25% discount as pure olive oil. Unfortunately, the reprocessing had also made it deadly. The chemical contaminant which actually caused the deaths has not been completely identified, but the unfortunate results are well known. If there can be any defense, it would be that the intent was economic fraud only. Surely, the firm had no intention of creating such a deadly situation.

Another example occurred in France in 1953. A baker received a shipment of flour made from rye. He could tell that the grain had been badly infested with the smut fungus because of its color and smell. He disregarded this, and the flour was mixed with wheat flour and baked into bread. The baker knew better than to sell the bread in his own village so

[1]An "action" level refers to the amount of a compound in food. Products having levels above the action level would be condemned by the FDA. The level is set by FDA after consideration of all available information including safety data described later in this chapter. A "tolerence" level is similar but requires a more formal procedure. Tolerance levels are usually more permanent.

he shipped it to another village. This incident resulted in 150 cases of severe illness and four deaths, caused by the alkaloid ergot (see Chapter 15).

Natural Compounds

There are literally hundreds of compounds found in natural products that have varying degrees of toxicity. There is relatively little quantitative data in this area, and it may well be the subject of much more study in the future.

Food Additives

The compounds added to foods for specific purposes are called food additives (see Chapter 3). Examples are antioxidants added to oils to prevent rancidity, preservatives to prevent food from spoiling, colorants to make food more appealing, etc. Most of these chemicals are generally regarded as necessary and safe.

Pesticides

Pesticide chemicals are used in the production of plants and animals. Many chemicals are essential for controlling insects and fungi in plants and to promote better health in animals. Residues may find their way into food. A well-studied example is DDT.

Environmental Pollutants

Environmental pollutants in land, air, and water can appear in food. Examples are lead from gasoline, cadmium from industrial metal plating, mercury in fish, from naturally occurring mercury in the ocean etc. One example that received considerable attention from the media concerned polychlorinated byphenyls (PCB). PCBs are a class of industrial chemicals that are very stable and virtually indestructible in the environment. They became commercially available in the United States around 1930, and were widely used in electrical transformers, heat transfer systems, hydraulic fluids, plastics, adhesives, sealants, paints, varnishes, insulating tape, fluorescent light starters, printing inks, pressure-sensitive copy paper and many others. This ubiquitous use, plus their exceptional stability, led to a build-up in the environment, which became apparent when the analyses used for DDT were modified to detect PCBs. Many so-called DDT levels were probably partially due to PCBs.

The first documented case of PCB contamination of food occurred in West Virginia in 1969, where FDA scientists found PCBs in milk samples. They traced it to used transformer fluid that had been added to a herbicide sprayed near cattla grazing areas.

Seven other cases were found, but the one which received the most publicity in the media was an industrial accident in Billings, Montana in 1979. About 200 gallons of PCBs at the Pierce Chemical Co. leaked from a damaged transformer into meat meal, which was subsequently sold for animal feed. This one incident required the destruction of over 300,000 chickens, 1,000,000 eggs, 75,000 frozen cakes, and 16,000 pounds of pork in Montana, Idaho, and Utah. Other incidents have occurred in Iowa, New York, North Carolina, and Puerto Rico. Some were traced to recycled carbonless paper, and silo sealants, but the worst cases were due to transformer fluid. When the magnitude of the problem becaue apparent, corrective action was taken and production of PCBs in the United States was stopped in 1977.

The health problems related to PCBs surfaced in the late 1930s as skin rashes, called chloracne, in industrial workers with high exposure to PCBs. In Japan in 1968, a substantial number of people ate rice oil grossly contaminated with PCBs and developed a number of adverse symptoms. Fortunately, even before the Japanese misfortune, the FDA was moving to protect the American public, and tolerance levels for PCBs in a number of foods were established. PCBs are a known carcinogen for animals, but there is no evidence that the low levels found in foods have any harmful effect in humans. Repeated epidemiological surveys have been negative. Yet, PCBs are everywhere. Surveys have indicated that one-half the U.S. population has detectable PCBs in their body tissues. The levels in the environment will gradually decrease, but the PCBs will be with us for some time.

All of the preceding situations involve a judgment of safety for compounds in the food supply and the environment. A judgment of safety may also be defined in the reverse sense as an estimate of risk.

THE RISK-BENEFIT CONCEPT

In the food safety area, perhaps no concept has stirred more debate than the risk–benefit concept. The questions are, "Risk to whom? and benefit to whom?" Also, there is the philosophical concept as whether one should indeed trade off an economic benefit for a health hazard. These are difficult questions.

Since the passage of the 1906 Food and Drugs Act, the FDA has taken the position that there are no harmful substances in food. Recently, they changed their opinion. It is obvious that there are harmful materials in food, and the questions are different. One should no longer ask "Are there harmful substances in food", but rather, "Are they significant?" Under the Biomedical Research and Training act of 1978, the National Cancer Institute is required to issue a comprehensive report on carcinogens in the environment. The data in these reports point out very clearly that absolute safety is unattainable, so we have to be able to use a system of evaluating the risks.

The Estimation of Risk

We can divide the estimation of risk into two major areas, those for which direct or hard data are available and those for which we have to use indirect, i.e., animal data. In the first category we can use death statistics. Fortunately, the United States has good death-rate statistics. We can divide the number of deaths by an estimate of the number of people exposed and come up with an estimate of risk. Table 13.1 lists risks from this type of calculation. We can do this for all types of hazards. For example, Table 13.2 lists risks from cancer. The range is very large and we are more willing to accept some risks than others. For example, many people particularly males, are willing to accept the risks from contraceptive pills.

One can express risk in many ways. Regardless of the method, an estimate of risk can be obtained from death statistics and exposure. With

Table 13.1. Annual Risk of Death

Activity	Risk per Year
Coal mining (black lung)	1/125
Coal mining (accidents)	1/770
Motorcycle racing	1/550
Horse racing	1/750
Automobile racing	1/830
Rock climbing	1/1,000
Fire fighters	1/1,250
Agriculture	1/1,700
Canoeing	1/2,500
Airline pilot	1/3,300
Motor vehicles	1/4,500
Football	1/25,000
Skiing	1/33,000
Drowning	1/53,000
Home accidents	1/83,000
Bicycling	1/100,000
Electrocution	1/200,000
Vaccination	1/330,000
Tornado	1/2,000,000
Lightning	1/2,500,000
Bee and wasp stings	1/10,000,000
Sharks	1/1,000,000,000

Table 13.2. Annual Risks from Cancer

Activity	Risk per Year
Cancer from sun (curable)	1/200
Cancer from sun (deaths)	1/16,000,000
Smoker (all effects)	1/300
Smoker (cancer only)	1/800
Contraceptive pills	1/50,000
Diagnostic X-rays	1/100,000
Black hair dye	1/16,000,000

safety problems in foods, however, this approach cannot be used simply because there are no death statistics. How can one calculate a risk from DDT, when there is not a single confirmed death attributed to recommended uses of DDT. There have been industrial accidents, but no deaths from recommended usages. In the situation, we have to rely on indirect data obtained by two approaches. One is human epidemiological studies, described later, and the other is the use of laboratory animals. Rats and mice are the usual animals, but sometimes rabbits, dogs, monkeys, fish, and other species are employed. The best estimates of human risk are obtained from animals that metabolize the compound in question in a manner similar to man.

Animal Models

Let us examine what happens when a food or a chemical is fed to animals. In Fig. 13.1, the dose is plotted against the animal response. The response could be weight loss, loss of hair, loss of appetite, tumor formation, etc. One of four types of responses can happen. (A) The compound has no effect. Starch, for example, in reasonable amounts, would be in this category. (B) There is no threshold and any amount will cause a response. Theoretically, there probably are few, if any, compounds in this category, but for conservative reasons, described later, this is the response usually adopted for compounds that cause cancer (carcinogens). Thiourea is an example. (C) There is a dosage below which there is no effect since the animal can detoxify or metabolize the compound. Above this level, the ability of the animal to handle the compound is overwhelmed and an adverse effect is seen. Many compounds are in this category, for example, the toxin which causes botulinum food poisoning. (D) There is an adverse effect due to too little of the compound, followed by an optimum level and then a level with an adverse effect. Some nutrients such as vitamin D and selenium are in this category, but the classic example is oxygen. The threshold level effect illustrated in (C) probably occurs most often.

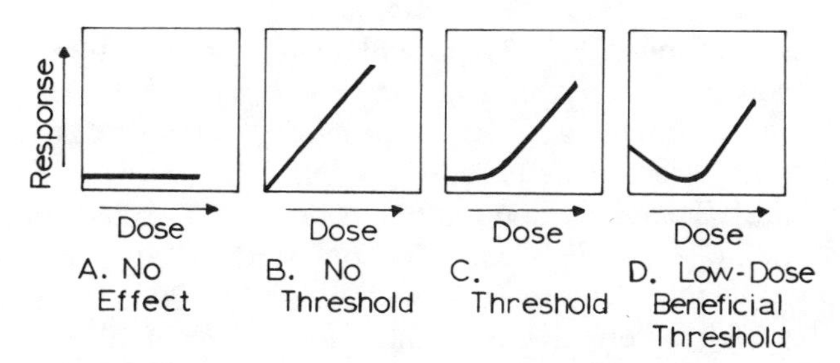

Fig. 13.1 Possible forms of the curves relating dosage to response. (A) No effect. (B) No threshold. (C) Threshold. (D) Low dose beneficial. Threshold.

The Threshold Level

Let us take a relatively simple case of a compound, which does not cause cancer, and which a company wished to introduce into the food supply. There are many types of information that have to be supplied to the Food and Drug Administration before the additive is approved. They involve chemical identity of the additive, degree of purity, method of analysis, source or method of synthesis, how it is absorbed or metabolized, function in the food, etc. Let us consider only the testing for safety.

Animals, usually rats and mice, are used for safety testing. Figure 13.2 represents an idealized experiment. Compound X is fed to groups of animals at increasing dosages as shown on the horizontal axis. A similar group, called a control group, is fed the same diet without compound X

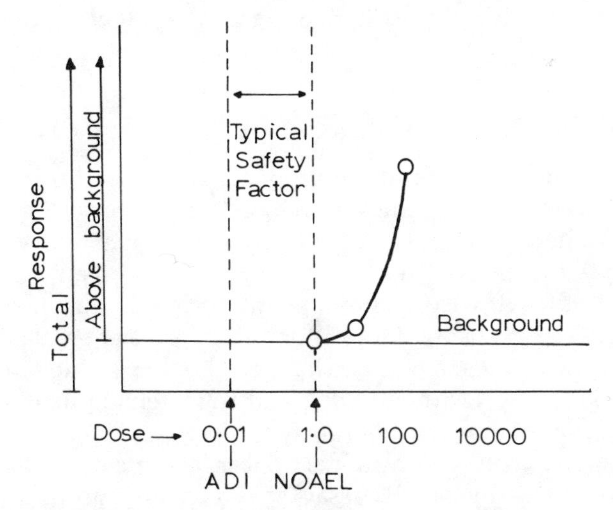

Fig. 13.2. A typical animal feeding experiment.

and a response is measured. There will always be an average or background response and anything above this can be attributed to compound X. There also will be a concentration of compound X below which there will be no effect. This is called the No Observed Adverse Effect Level (NOAEL). This level divided by 100 is usually called the Acceptable Daily Intake (ADI). It is usually expressed as milligrams per kilogram of body weight. The reason for dividing by 100 is, first, that the animals may be less sensitive than man by a factor of 10. Second, humans may vary in sensitivity by a factor of 10. The two multiplied together give 100, and this is the basis for the 100-fold safety factor. The ADI would probably be the level allowed by the FDA for addition to food. Note that this is a conservative level. Since compound X is unlikely to be added to every food, human consumption is likely to be much lower. The above procedure is an oversimplification since data on at least two types of animals, one of which should not be a rodent, would probably be required. Also tests would be required over several generations, for varying lengths of time. The degree of complexity of the testing required would vary depending on (1) the chemical structure of the compound, (2) whether it was found widespread in nature, (3) what the estimated comsumption level would be, (4) whether it was intended as food for babies, and many other factors. These procedures are well understood and are accepted worldwide. They have worked very well.

The above model is only appropriate for compounds to be added to food. It does not apply to compounds found naturally in foods. For example, solanine, an alkaloid found in potatoes, may be present, not at 1/100, but at 1/7 of the NOAEL. This situation is unavoidable in our food supply. Yet there is no cause for concern—the safety factors are quite adequate.

Safety Testing for Carcinogens

The preceding model is not appropriate for safety testing for compounds suspected of causing cancer. First, many texicologists do not believe that there is a threshold value for some carcinogens. This may mean that the threshold level is so low that we really cannot economically test for it. Also, many human cancers do not show up until 20–30 years after exposure. More important, the Delaney Clause in food safety legislation states that no carcinogens should be added to our food. The 100-fold safety factor just does not offer enough protection for chemicals suspected of causing cancer.

In view of the above, the National Cancer Institute in the 1960s developed a procedure (the NCI Bioassay) to screen large numbers of chemicals for carcinogenicity. This program is currently operated by the National Toxicology Program (NTP). A typical test would use two species of animals (usually rats and mice), with 50 animals of each sex in each of a control group, a "low" dose, and a "high" dose group. This makes a total of 600 animals. After 2 years, the animals are sacrificed and thousands of microscope slides are prepared from about 40 of the organs of the animals. The slides are examined for evidence of tumor formation. Obviously, such testing takes time and money. A typical experiment involving one chemical costs about $600,000. Obviously there is a limit to the number of chemicals that can be screened.

The Risk Concept

The NTP procedures for screening compounds for carcinogenicity are working well, but they may be getting more publicity than was originally intended. The procedures were designed for screening purposes, but some of the data are being used to calculate risks. After all, it is not enough to know that a compound may cause cancer. We really would like to know what the risk is. Ideally, the testing should be repeated with more test levels. But this is very expensive. One experiment on saccharin alone cost over $1,000,000. In view of this, there has been a tendency to use the available data to calculate a risk to humans. Because these are only two test levels, the model in Fig. 13.1B is usually adopted, since it gives conservative conclusions. This is justified, since, after all, where human safety is concerned, one should err on the side of safety.

There are problems with this type of approach, however, as illustrated in Fig. 13.3. Suppose tumor formation was observed at dose C, and the actual response curve was the solid line. Tumor formation at doses A and B would be much less than those calculated if the linear dotted line response were employed. Some of these differences can be very large, since often 100-fold extrapolations are employed from the high doses actually used to the low doses expected to be encountered in the real world. Increasing the number of animals would get statistically signiifcant results for lower dosages, but this is practically and economically not feasible. The net result of using high dosages and extrapolating to low doses is that a large uncertainty is introduced into the degree of risk. And, there is still another problem. The data for animals must be interpreted for humans, and this introduces a second uncertainty. There are statistical methods for extrapolating from high to low dosages and for interpreting animal data for humans, but the methods may give estimates of risk in which one may be 200,000 times as much as another.

This uncertainty in extrapolation may create another kind of problem.

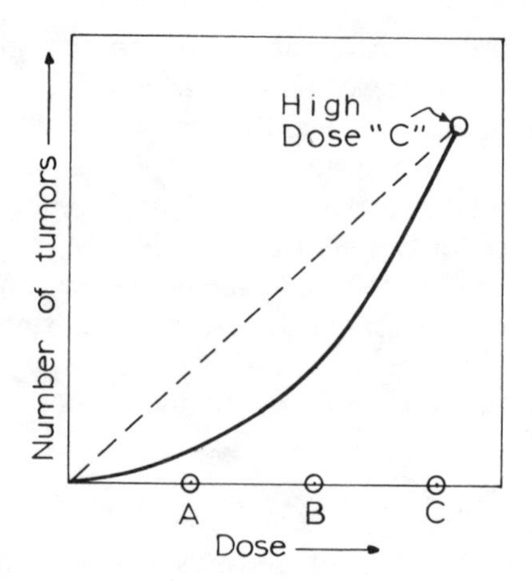

Fig. 13.3. The difference in estimation of risk using a linear and a nonlinear response curve.

If we extrapolate the levels fed to animals and to get equivalent human food intakes, we may get a situation that does strain ones credibility. This "box car" approach has been used to ridicule the animal testing data. Admittedly, the amounts of foods necessary to ingest the additive levels equivalent to that fed to animals per day (DES: 3000 tons of beef; calcium cyclamate: 350 12 oz. bottles of soda) would strain almost anyone's capacity. Therefore, one should be careful in extrapolating data even though it is one of the most important ways of getting an estimate of risk. It does point out, however, that the estimates of risk must be viewed with caution. Fortunately, as the sciences of toxicology, biochemistry, and human metabolism evolve, the way in which the human body handles a particular compound becomes better known. This, in turn, leads to a better definition of the shape of the response curve, leading to a better choice of statistical methods for extrapolation and interpretation. This will lead to better estimates of risk.

Epidemiological Studies

Epidemiology has the real advantage of actually identifying human risk factors. In a typical experiment, say, exposure to an industrial chemical, a group of workers exposed to the chemical are identified. A group, similar in all other respects, but without exposure to the chemical, is located, and the incidence of the response in question is compared for the two groups. This approach has been very useful in determining human risk. The International Agency for Research on Cancer (IARC) has reported 18 chemicals and industrial processes that have been found to cause cancer. This approach can also be used to validate the estimates of risk to humans from animal data, which are not always transferrable. For example, laboratory studies confirmed that large doses of saccharin produced bladder tumors in the second generation of male rats, but epidemiological studies on human diabetics showed no increase in bladder tumors.

Recent studies also showed that mice do not develop bladder tumors and actually showed a decrease in liver tumors compared to the controls. This raises the interesting question as to whether humans are more like rats or mice.

Epidemiology was used to study the risks of coronary heart disease. Studies on both the incidence of heart disease and the mortality have led to the discovery of a number of risk factors. These are heavy smoking, high blood total cholesterol level, high blood pressure, diabetes, high blood triglyceride levels, low plasma high density lipoprotein, obesity, stress, lack of exercise, and personality type. The risks involved the first three are listed in Table 13.3.

Another example of the epidemiological approach concerns the risks of smoking. Figure 13.4 illustrates the correlation between cigarette smoking and lung cancer and it is striking. The rate for women is much lower than that for men because they have been smoking for a shorter period. The scientists expect that the lung cancer rate for women will equal that of men in another 25 years. Some have even suggested that the longer life expectancy of women in the United States is primarily due to the difference in smoking habits. In addition, there is an increased risk of other cancers for smokers (Fig. 13.5). For example, smokers have ten times the risk of getting cancer of the pharynx.

The consensus approach is gaining favor in many of the controversial topics in food safety and public health today. Examples are the role of salts in hypertension, cholesterol and its derivatives in heart disease, lipids and fiber in several diseases, and the list goes on.

Table 13.3. The Effect of the Major Risk Factors on the Incidence of the First Major Coronary Event

Risk Factor[a]	Incidence per 1,000 population
None	23
Any one	54
Any two	97
All three	189

[a] The risk factors were heavy smoking, elevated blood total cholesterol, and elevated blood pressure.

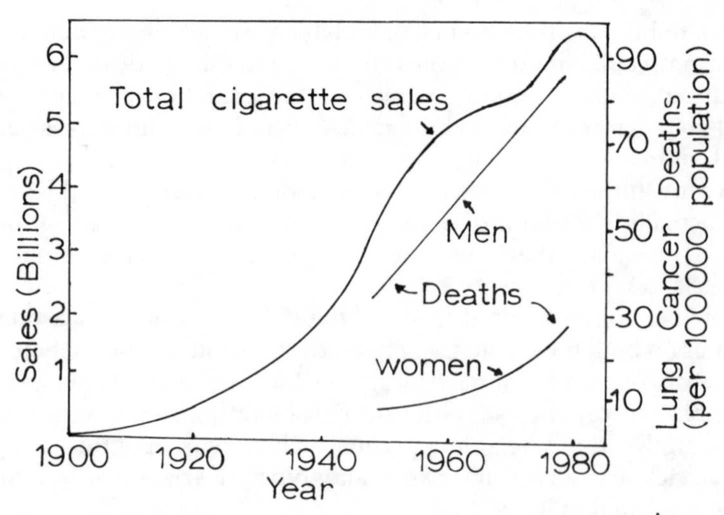

Fig. 13.4. Relationship between cigarette sales and deaths from lung cancer.

Epidemiological studies can be used for a wide variety of purposes. For, example, one recent study listed 40 risk factors for human cancer. They included sexual promiscuity, obesity, age at first pregnancy, exposure to some diseases, etc. Such studies provide excellent suggestions for research on possible causes of cancer. Yet the epidemiological ap-

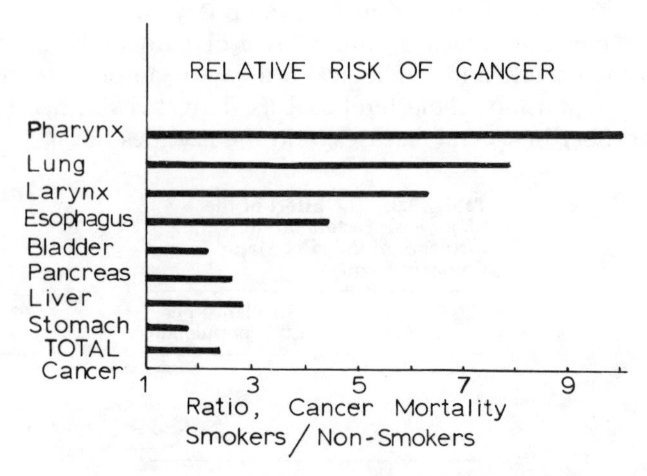

Fig. 13.5. Comparison of the mortality from cancers in a number of sites for smokers and non-smokers.

proach has its own problems. For example, if one wants to study the effect of caffeine in coffee, how do you locate a large control group with no exposure to coffee? Some investigators have commented that the errors involved in epidemological studies may be large enough to hide a minor effect. This may be true, but the overall approach has been very useful and very successful.

Other Approaches

A wide variety of approaches have been developed to help answer the question: Does this compound produce cancer? Perhaps the best known is the group of tests called the Ames Test developed by the microbiologist Dr. Bruce Ames. These tests involve a series of microorganisms with specific biochemical requirements, which can be used to predict whether a compound is a mutagen. A mutagen is a compound that can cause a change in an organism that is maintained in following generations. The new organism is called a mutant. Presumably, the ability to cause mutations is correlated with the ability to cause cancer, so these tests are very useful for screening purposes.

Many microorganisms, including protozoa, as well as a number of enzymatic reactions, have been suggested as potential screening methods. When these tests are combined with animal data, epidemiological data, and knowledge of the metabolic route of a compound in man, the estimates of risk to humans are much better.

RELATIONSHIP OF ANIMAL DATA TO HUMANS

The question is often raised "How well do the tests on rats and mice predict the ability of these substances to cause cancer in humans? What we do know is, all but one of the chemicals known to cause cancer in humans, cause cancer in animals. The exception is arsenic. There are seven cases so far in which chemicals were first found to be carcinogenic for man. There are several hundred chemicals found to be carcinogenic in animals which, so far, have not been found to be carcinogenic in man. If a chemical is carcinogenic in several species of animals, it is likely to be carcinogenic in man. An example is aflatoxin, produced by a mold growing primarily on peanuts and corn. It is carcinogenic in mice, rats, fish, ducks, turkeys, marmosets, tree shrews, and monkeys as well as humans. Other examples are asbestos, diethylstilbestrol (DES), and benzidine. However, many chemicals are carcinogenic in just one species. Of 98 chemicals found to be carcinogenic by the National Cancer Institute in rats or mice, 54 were positive in only one of the two species. It is obvious from this background, that in order to judge human risks, animal species should be chosen which metabolize the compound in the same way as humans. Everyone agrees that our society should make every effort to reduce the occurrence of cancer. Yet our efforts should be directed towards situations that have real risks instead of wasting our resources on situations with insignificant or even imaginary risks. The only practical way to approach this problem is to refine the research and methodology to produce realistic estimates of risk. This will have to be followed by estimates of acceptable risk. Current thinking in the scientific community defines "acceptable risk" as one extra death per million over one's entire lifetime. This seems to be reasonable and the legislative and scientific community should be encouraged to make it realizable.

The background data on which to establish estimates of food safety and subsequent risk to humans for food additives are well-developed. Data for safety testing on natural components of foods are less well-developed and probably will be an area of increased research activity in the future. The estimation of risk to humans from compounds that may cause cancer is probably the least developed, though it is the object of a great deal of current research. This area is certainly the biggest and the most important.

ESTIMATES OF BENEFIT

If we think that the estimates of risk are fraught with uncertainties, the estimates of benefit are much worse. First, they can only be done with difficulty on compounds that have been in the environment for some time. Estimation of benefit for compounds to be introduced is much more tenuous. Regulatory authorities are, understandably, not very enthusiastic about having to make estimates of benefit for compounds in the environment. Consideration of benefits are not required in current considerations of food safety, but if and when they are, they will obviously require value judgments. The judgments will need a consensus approach in which observations from every area will be needed to form optimum decisions.

Michael Jacobson, of the Centre for Science in the Public Interest, a Washington-based consumer group, has advocated a common-sense approach to risks. He divided them into three categories:

1. Serious risks: one in 5,000 for chances of harm. Change your living and buying habits.

2. Smaller risks: between one in 5,000 and one in 5 million. Support legislative action to reduce risks. Personal changes are optional.
3. Trivial risks: less than one in 5 million. Don't worry about them.

BIBLIOGRAPHY

ANON. 1973. Toxicants Occurring Naturally in Foods. Committee on Food Protection. National Academy of Sciences, Washington, DC.

ANON. 1981. Regulation of Potential Carcinogens in the Food Supply: The Delany Clause. Report No. 89. Council for Agricultural Science and Technology, Ames, Iowa.

ANON. 1981. The Risk/Benefit Concept as Aplied to Food. A Scientific Status Summary by the Institute of Food Technologists, Chicago, Illinois.

HAVENDER, W. R. 1984. Of Mice and Men. Report by the American Council on Science and Health, New York.

LAWRENCE, N. N. 1976. Of Acceptable Risk: Science and the Determination of Safety. William Kaufman Inc., Los Altos, California.

STERRETT, F. S. and ROSENBERG, B. L. 1982. Science and Public Policy 11. New York Academy of Sciences, New York.

14

Food vs. Fuel

The long lines of motorists waiting for gasoline in the 1970s sent shock waves through the American public. In the United States big cars were part of the national way of life. We were used to driving wherever and whenever we wanted. Gasoline was cheap. The rest of the world taxed gasoline very heavily, thereby encouraging the development of small fuel-efficient cars. In the early 1970s, a group of countries formed a cartel called the "Organization of Petroleum Exporting Countries" (OPEC), and raised the price of crude oil by a factor of 10-15. The search for alternative fuels began in earnest. One of the solutions was "gasohol," a blend of 10% ethyl alcohol (ethanol) and 90% gasoline. The ethanol could be produced from the large surpluses of corn available in the United States. In 1980, Congress passed legislation calling for production of 500 million gallons of ethanol for fuel in 1981, one billion gallons in 1982, rising to 10 billion gallons in 1990. The 1990 goal would have required about 100 million tons of corn, which is slightly less than half of the 1982 corn crop (210 million tons). This did not happen and, and we may be thankful that it didn't, since it would have raised the cost of food.

THE RISE AND FALL OF GASOHOL

The 1980s saw a burst of promotion for gasohol. Part of the reason for the popularity of gasohol was a 5-cent per gallon federal tax exemption on gasoline plus exemption from state taxes in many areas. The two exemptions could add up to as much as $39 per barrel for the ethanol portion of gasoline. The public enthusiasm was short-lived, however, and sales were soon discontinued except in the corn-producing areas. Gasohol developed a poor public image for several reasons. Some operators may have added more than 10% ethanol, which caused engine performance problems. Possbily there was some moisture absorption in the transportation system causing corrosion and poor engine starting. Regardless of the reasons, the term gasohol has almost disappeared, but ethanol is still being used in increasing quantities as an anti-knock compound in "super-unleaded" gasoline.

The prospect of two markets for corn and other carbohydrate crops made the food supply economists very nervous. Never before had there been sizable alternate markets for food. In 1980, gasoline was over $1 per gallon, whereas in many European countries, the price was three or more times as high. If the price of crude oil had continued to rise, as some predicted, it would divert food to fuel in large quantities, and the price of food would rise. The poorer countries would be unable to afford the price of food on the world market. Fortunately, in the mid-1980s the price of gasoline has come down and does not impact directly on the price of food. However, this may be a temporary respite.

THE ECONOMICS OF ETHANOL PRODUCTION

The economics of ethanol production are both interesting and confusing. There are apparently several levels. For example, the Archer-Daniel-Midland Company, the largest producer of fuel ethanol in the United States gets the following from one bushel of corn: 9.2 lb of 21% feed protein, 2.7 lb of 60% gluten meal, 3.5 lb of corn germ, and 31.5 lb of starch. Only the starch would be used for ethanol production, and all the others are saleable agricultural products. A less sophisticated production system produces 20.5 lb (2.6 gal) of ethanol, 20 lb of carbon dioxide, and 17.8 lb of Distillers Dried Grains (DDG) from one bushel of corn. The carbon dioxide could be sent to the oil companies for use as an aid in the production of crude oil. The DDG could be sold as a protein concentrate for animal feeding. DDG contains about 27% protein, so 1 pound of soy meal is equal to about 1.63 pounds of DDG. A third process could be a combination of both feed and alcohol. The economics of this are interesting (Fig. 14.1). The DDG from 20 bushels of corn added to 80 bushels as feed provided more gain in weight than the 100 bushels of corn. Part of the reason for this is that the DDG was enriched in protein, above that in the original corn, by the cells of the yeast from the fermentation step. The lower amount of carbohydrate fed to the animals could be made up from inedible (to humans) roughage such as corn stalks.

One of the major concerns of the food supply economists about the development of the fuel from food industry is that it would compete

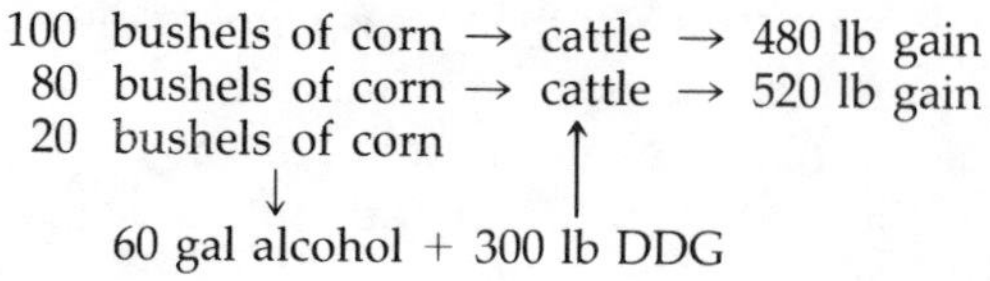

100 bushels of corn → cattle → 480 lb gain
80 bushels of corn → cattle → 520 lb gain
20 bushels of corn ↑
↓
60 gal alcohol + 300 lb DDG

Fig. 14.1. Corn for feed and alcohol

with supplies of human food. In the United States, over 85% of the corn crop either goes to animal feed or is exported for animal feed. The ethanol from the corn program would not compete with human food, but it has been suggested that it might impact on the animal feed situation in two ways: calorie supply would decrease but protein concentrate supply would increase. The 1984 production of fuel alcohol is estimated at about 170 million gallons of ethanol corresponding to about 4.2 billion gallons of super unleaded gasoline. About 39,000 tons of DDG would be produced as a by-product of ethanol production, and this is equivalent in protein content to only about 29,000 tons of soy meal. Since the 1984 production of soy meal was about 30 million tons, the ethanol production system would have a minor, although positive, impact on the supply of protein concentrations for animal feeding.

ECONOMICS OF ALCOHOL IN GASOLINE

It is obvious that addition of ethanol to gasoline was a short-term practical possibility to extend the available supplies of motor fuel. This was successful for the short term, and was encouraged by government subsidies, but it is unlikely to be successful on a long term basis.

Table 14.1 shows the energy value of ethanol and methanol compared to gasoline. Obviously, both are lower than gasoline and, as a means of diluting gasoline, would make a more expensive mixture (Table 14.2). If cost alone was the deciding factor, ethanol would not be used, but ethanol is added because it raises the octane value. As engines have

Table 14.1. Energy Value of Fuels

Fuel	Energy (BTU/gal)
Gasoline	128,000
Ethanol	82,000
Methanol	64,000

Table 14.2. Octane Value and Cost of Blending Agents for Gasoline in 1984

Additive	Octane value	Cost (cents/gal)
1. Methanol	115	0.43
2. Ethanol	112	1.80
3. Methyl *t*-butyl ether	106	1.00
4. Toluene	106	1.03
5. *t*-Butyl alcohol	99	1.00
6. *n*-Butane	92	0.52

become more sophisticated, the compression of gaseous fuel in the cylinders has increased. The increased pressures caused the fuels to burn prematurely. This causes a "knock" in the engine, so fuels with higher octane ratings are needed. Octane is the reference compound; a gasoline mixture with an octane rating of 100 is equal to pure octane. Formerly, the most common antiknock compound in gasoline was tetraethyl lead, but since lead is a toxic polluter of the environment, the Environmental Protection Administration (EPA) has legislated the removal of lead from gasoline. The banning of lead compounds has led to a demand for other compounds to raise the octane rating of gasoline. Ethanol and methanol can both be used to raise the octane rating of gasoline.

THE PROALCOOL PROGRAM IN BRAZIL

The rise in oil prices in the 1970s affected all consumers, but some countries were hit harder than others. Brazil in particular imported a great deal of oil and decided to develop an alternative fuel. Brazil is the world's largest producer of sugar, and since their potential to produce sugar far exceeds their consumption, they chose to make alcohol the fuel for automobiles. They set a goal for production of 2.4 billion liters of ethanol in 1978-1979, enough to replace 20% of the gasoline estimated for use in 1980. The actual amounts of ethanol produced are shown in Table 14.3. They are right on target. The Brazilians have two major potential crops for ethanol production, sugarcane and manioc (cassava). They chose sugarcane because of the horticultural history of large scale production. It is also a very efficient crop. By way of comparison, an acre of sugarcane will yield 700–1,000 gallons of ethanol, as compared to 400–800 gallons from sweet sorghum, and 300–400 gallons from corn.

Two types of fuel for automobiles are available in Brazil. One is a mixture of 20% ethanol in gasoline, and the other is 100% ethanol. The

Table 14.3. Ethanol Production in Brazil[a]

Year	Ethanol (million gallons)
1980	680
1983	1780
1985	2360
1990(est)	3150

[a] The 1983 gasoline consumption is approximately 3 billion gallons. The 1985 ethanol goal is equivalent to 9 million tons of sugar or 25 million tons of corn.

government is encouraging the use of ethanol (Alcool in Brazil) in a variety of ways. The price of 100% ethanol is set at 65% of the cost of the 20:80 gasoline. The road tax is $60 for Alcool cars and $120 for gasoline cars. Gasoline pumps are closed on both Saturdays and Sundays, whereas Alcool pumps are closed only on Sundays. Loans can be obtained on more favorable terms for Alcool cars, etc. Ethanol cars have had problems such as corrosion, engine wear, and difficulty in starting, but engine makers have solved the engine starting problems by starting the engines on gasoline and switching to ethanol. They say the corrosion problems have been overcome by changes in engine design and contraction. The ethanol fuel in Brazil is still more expensive to produce than gasoline, but the legislators hope that the price will come down as distilleries become more efficient. There is no question but that the ethanol program has succeeded, and the authorities have just announced their intention of converting agricultural tractors to use ethanol. One of us (F.J.F.) had the privilege of spending 4 months in Brazil and seeing this program firsthand.

The Brazilian experiment is being watched carefully by other countries in the world. The choice of raw material will depend on availability in a particular country. The United States is using corn, whereas Australia is seriously considering wheat. Malaysia and the Philippines conceivably could use manioc. The success or failure of any of these programs will depend on the price of cereals, the cost of processing to produce ethanol, and the price and availability of crude oil. Curiously, in 1963, there was an estimated 36 year supply of crude oil. In 1983, again there was an estimated 36 year supply. In 1984, there was an oil glut on the world market, and this caused a cooling-off of the plans for alternate fuels. At the same time, there was a glut of sugar on the world market, and the price reached an all-time low of 3.7¢ per pound. This is estimated to be only one-third of the cost of producing the sugar, and countries that depend mainly on sugar exports for foreign exchange are in deep recession. Sugar production apparently does not respond quickly to the laws of supply and demand, and many producers simply increased production to make up for lower prices. Prices as low as those in 1984 are a real incentive to convert part of the production to alcohol; however, most of the cost of the ethanol is in the distillation process, not in the cost of the raw sugarcane.

THE POTENTIAL FOR METHANOL

In the United States, and possibly worldwide, methanol is likely to replace ethanol as a fuel for automobiles, primarily because it is cheaper (Table 14.2). Methanol has a lower energy content (Table 14.1), but it is the preferred fuel for racing cars. It is the main component in "Dry-Gas," used to prevent water from freezing in the gasoline lines in autos. Methanol is readily formed from methane gas, coal, wood, or almost any type of biomass. It is a well-known industrial chemical produced cheaply in very large quantities. So why is it not used for auto fuel? The answer seems to be a "chicken and egg" situation. Methanol requires a redesign of the automobile engine, and auto makers are reluctant to produce methanol cars until an assured supply of fuel is available. Methanol suppliers are reluctant to build mammoth methanol plants without some assurance that the engines will be available. However, there still is considerable interest in methanol cars. The State of California had a fleet of 500 cars running on methanol in 1983. The Bank of America had a fleet of 292. These are experimental programs, and they have demonstrated the superiority of methanol as a fuel.

Another use has been found for methanol. It is being used to raise octane ratings in conventional gasoline, because of its obvious price advantage over ethanol. However, it has the disadvantage that a co-solvent such as methyl *t*-butyl ether, toluene, *t*-butyl alcohol, or *n*-butane must be added, otherwise, the methanol tends to absorb moisture from the air and separate from the gasoline. These co-solvents are slightly higher than the current price for gasoline, and there may be a question about their supply. They all come from petroleum and therefore would be related to the cost and availability of petroleum.

Methanol will probably replace ethanol eventually even in countries like Brazil. Climates that enable sugar cane to grow efficiently are also favorable to fast-growing trees such as Eucalyptus. A Eucalyptus tree will grow to a diameter of 6 inches in 7 years. Current research in Brazil has demonstrated that yields up to 61 tons of dry matter per hectare are available currently and they hope to obtain 100. The wood can be used to make methanol. Natural gas is used today as the raw material to produce industrial methanol, but it is anticipated that coal will be the source in the future. Coal is likely to be cheaper than any source of biomass in the near future, but the economics are hazy and difficult to predict.

Prediction for the future of any biomass energy application will depend on a number of interlocking considerations. The energy considerations fall into two groups–large generating stations for industrial and home heating, and portable fuels for transportation and possibly some home heating. The former has and will be dominated by coal, hydropower, nuclear power, and possibly natural gas. Nuclear power is destined for a larger role and may even eventually dominate the others. France, for example, in 1984, produced nearly 60% of its power from nuclear stations. On the other hand, countries like Brazil, China, and Canada still have potential sites for hydropower and probably will develop them.

The portable fuel category is much more complicated. A highly concentrated energy source is needed for autos, planes, trains, trucks, etc., and it is likely to remain gasoline and diesel oil, obtained from petroleum, for at least the next two decades. The reason is simple–petroleum is cheaper. Eventually, as the supply of readily obtainable crude oil diminishes, several other sources will have to be phased in. These will be liquid fuel from coal, tar sands, oil shale, and viscous petroleum deposits. The economics will depend on availability as well as many other factors. South Africa already obtains a large proportion of its liquid fuel requirements (gasoline) from coal by the well known Fischer–Tropsch process. Obtaining oil from tar sands is a reality in Canada and will become economically more promising when the price of crude oil rises again. The production of oil from oil shale has been demonstrated to be technically feasible in the Unted Sates, but the economics today are unfavorable.

The production of fuel from biomass is even further distant than that from fossil fuels, but there is no dearth of ideas. The concept that has received the most media attention is probably alcohol from cellulose. The cellulose could come from a variety of sources, including newspapers, corn cobs, trees, sawdust, and other waste products of the agricultural industries. Even sewage is a viable contender. Converting newspapers to alcohol requires a two-stage process. The first is an acid or enzyme hydrolysis to break down the cellulose, followed by a fermentation to produce ethanol. The processes involving sawdust, trees, etc, also require an additional step to remove the lignin. Lignin is a three-dimensional cement that binds the cellulose fraction together. It is resistant to both acid and enzyme hydrolysis, and older processes degraded the lignin. Recent research has developed ways, such as solution in phenol, to remove the lignin more or less intact. It can then be utilized in the production of adhesives, and the additional income from lignin may make the whole process more economical.

The utilization of whole trees in leaf has some exciting possibilities. In addition to alcohol and lignin, the residue contains considerable protein and fiber and is suitable for cattle feed. Trees can be grown economically on marginal soil and appear to have attractive commercial possibilities. For example, young poplar trees are already being grown for cattle feed. In Pennsylvania, hybrid poplar trees, cropped on a 4-year cycle, yielded 8 metric tons of dry matter per acre per year. In Oregon, annual cropping of aspen trees for fuel produced 25 tons of dry matter per acre. However, if the harvesting of whole trees becomes popular on marginal soil, fertility levels will soon decline and some form of artificial fertilization will become necessary. However, this may be a small portion of the overall cost.

There are a number of fascinating schemes to use the products derived from biomass as fuel and many other industrial uses. Dr. Melvin Calvin of the University of California found a species of tree (*Cobaifera longsdorfii*) in Brazil that produces virtually pure diesel oil. The natives have known about it for many years. They drill a 5- centimeter hole in the 1-meter thick trunk and put a bung in the hole. Every 6 months or so, they remove the bung and collect 15 to 20 liters of a hydrocarbon liquid. The liquid can be placed directly in the fuel tank of a diesel-powered car. The *Cobaifera* tree is merely one example of a wide variety of species that produce hydrocarbons. Calvin's chief interest for the U.S. is the genus *Euphorbia,* which produces a milk-like emulsion of hydrocarbon in water. On his northern California ranch, he has a stand of *Euphorbia lathyrus* that produces the equivalent of 10 barrels of petroleum per acre. He anticipates that the yield could be improved dramatically by breeding and genetic selection and cites natural rubber as an example. Research in Brazil has demonstrated the technical feasibility of using oil from avocados in diesel engines. Avocados grow wild in Brazil and have up to 25% oil. Research in several other parts of the world has shown that peanut and sunflower seed oil could also be used for diesel fuel. Actually, almost any type of oil could conceivably be burned as diesel fuel with the proper engine modifications, but the economics may be prohibitive.

It is likely that products from biomass may gradually replace some of the petroleum feedstocks in the future depending on technical feasibility, availability, and cost. For example, the hydrocarbons from *Euphorbia* react similarly to naphtha when processed, and naphtha is an important product in the chemical industry. We may see products from biomass competing for land utilized for food crops.

In some of the lesser developed countries, the production of "biogas" (principally methane) from agricultural wastes and sewage is well established. It is estimated that the Peoples Republic of China alone has 7.5 million biogas units. India expects to have over one million units. The use of biogas will lessen the use of wood as a fuel for cooking. In many places, wood is in short supply and an alternative fuel would be a great benefit. The biogas concept is being developed on an industrial scale in Oklahama where a company using the Calorific Recovery Anaerobic Process (CRAP) processes manure from the cattle feeder lots into protein concentrates, methane, and fertilizer. CRAP can handle one million pounds of manure per day.

One of the most publicized alternative energy sources in the United States is solar power. It is already widely used for partial heating of houses, greenhouses, swimming pools, etc., through a wide variety of collector devices. Government subsidies have contributed to the acceptability of solar units, but there is a question as to whether solar power is cost effective in these applications without a subsidy. Commercially, several projects have been attempted. The largest installation to use photovoltaic cells as collector devices to transform light directly into electricity is the 1000-kilowatt (kW) installation near Hesperia, California. A large installation was built in Saudi Arabia in 1982 that uses 41,000 photovoltaic cells to produce 350 kW of direct current. The current is fed into giant lead-acid batteries to even out the supply, and then to converters to produce alternating current for the villages. When the cost of the photovoltaic cells is included, the cost of the electricity may be as much as ten times that of electricity produced from their oil, but the Saudis expect that sunlight will be there long after the oil is exhausted.

The world's largest solar power unit is located in the Mojave desert in California. The unit consists of a boiler on a tower 300 feet above the desert floor, surrounded by 21,816 mirrors to collect sunlight from approximately 80 acres. The mirrors are computer controlled to focus sunlight on the boiler, which delivers steam at 950°F to a heat storage unit and then to a conventional turbine and electrical generator. The system has a maximum capacity of 12.5 megawatts (MW) of electricity and has functioned surprisingly well. Five other smaller units (two in Spain, one each in France, Italy, and Japan) are operational and an even bigger one (15 MW) is being built in California. This concept appears to have exciting possibilities.

Windmills were a familiar feature of the American plains many years ago and were used to generate small amounts of energy primarily for pumping water. They were displaced by diesel generators and rural electrification, but today wind power is making a comeback. In the Altamont Pass in California, two companies alone have 650 windmills generating over 40,000 kW of electricity. The power is sold to Pacific Gas and Electric Co., which has plans to expand their purchases of wind power to 138,000 kW. However, this amount is only 0.008% of the companies' steam, hydro, geothermal, and nuclear capacity.

All of these sources of renewable energy (biomass, solar, wind, geothermal, tides, waves, ocean temperature differentials, etc., with the exception of hydropower) comprise only a very small percentage of our energy use. But they all contribute and the potential for better use of them exists.

CONCLUSIONS

It is possible, but very unlikely, that potential sources of food or feed will become major sources of fuel in the foreseable future. We are confident that other sources of energy will always be cheaper.

BIBLIOGRAPHY

ANON. 1980. Food vs Fuel: Competing uses for cropland. Environment 22 (4),33–40.

ANON. 1984. Methanol touted as best alternate fuel for gasoline. Chem. Eng. News, June, pp. 14–15.

BUNGAY, H. R. 1982. Biomass refining. Science *218* (4573), 643–646.

GOLDENBERG, J. 1984. Energy problems in Latin America. Science *223* (4643), 1357–1362.

MORGAN, R. P., and SHULTZ, E. B. 1981. Fuels and chemicals from Novel Seed Oils Chem. Eng. News. Sept., pp. 69–77.

15
Potential for Disaster

PLANT DISEASES

Every person born on earth hopes to have enough to eat. For simplicity, let us assume that food means calories, and 2500 calories a day is a liberal allotment. Nearly all of the calories have to come from agricultural production since there are no other significant sources. Agricultural production means both plant and animal products but since animals feed on plants, the ultimate source is plants. Plants have always been subject to diseases and a number of well-documented disasters have taken place. The potential for disaster may be even greater today in spite of human ingenuity, because of the pressure on the land for increased yield and because of the increase in monoculture. The never-ending battle between plant pests and agricultural production will grow more intense. It will require all the resources of sophisticated agricultural technology to maintain a favorable balance.

Every food crop has its own spectrum of pests. A few examples are described in the following pages.

The Potato Blight

The Irish potato famine has the dubious honor of being perhaps the worst crop failure in history. It caused untold hardship in Ireland and triggered large waves of emigration.

Ireland in the eighteenth century was a poor country dominated by English absentee landlords who exacted a toll from the villagers and allowed them very small plots (a half-acre) of land to live on. The Irish learned that white potatoes could be grown in "lazybeds" raised several feet above the bogs. These beds, 6 to 8 feet wide and 100 feet long, could be located out of sight of the English soldiers and provided an excellent supply of food. The potato became a staple and sometimes was the only food in the diet. This dependence increased until disaster struck in 1845.

The spring of 1845 was warm and pleasant, but in the summer the temperature dropped as much as seven degrees below normal and cold rains began—weather ideal for the spread of the potato blight. The blight was caused by the growth of a fungus, *Phytophtora infestans,* which had overwintered on cull potatoes piled around the houses. A spore germinates on the leaves of potato plants and immediately spreads within the plant. In a matter of days, each germinating spore produces millions of additional spores and the disease spreads like wildfire. Most plants are killed, and any that escape with a mild infection may rot in storage or serve as sources of infection for the next year.

During the time the potato was the main source of food in Ireland the population increased from 4 million in 1800 to 8 million in 1845. After the failure of the potato crop in 1845, the land could not support that many people. Food aid from England enabled the Irish to survive the winter of 1845, but with the planting season of 1846 came the rains, providing optimum conditions for the growth of the fungus, and again the potato crop was destroyed. This time, the reserves had been used up and famine struck in earnest. The folklore of Ireland is filled with tales of horror of this period. Weakened by starvation, the people fell easy prey to dysentery, typhus, and many other diseases. In 10 years, the population of Ireland decreased by 3 million people; 1 million died and 2 million emigrated. They came to the United States, Canada, and other parts of the British Empire.

The early emigrants in 1845 and 1846 arrived in the New World prior to the outbreaks of disease. Later emigrants were not as fortunate, and the conditions aboard the ships for people already weakened with disease were awful. The conditions were similar to those on the slave ships. There was no place to cook, improper latrines, and certainly no place to heal the sick. On the journey across the Atlantic Ocean, one of every twenty Irish refugees died and was buried at sea. One can imagine their condition when they arrived in the United States and Canada. Local residents were not happy to see emigrants arriving on disease-ridden ships. A monument to the Irish refugees still stands in Montreal—a mound containing the bodies of 6,000 Irish. In spite of the problems, a tremendous number of Irish were absorbed by the United States and Canada. They have left their mark. The railroads across Canada in the 1800s were built by gangs of Irish "navvies." Three families, the Fitzgeralds, the Kennedys and the Reagans produced American presidents.

Ireland was not the only land to suffer the effects of the potato blight. The failure of the potato crop in 1916 in Germany was a decisive factor in the inability of Germany to win World War I. Seven hundred thousand Germans starved to death during the winter of 1916–1917. In 1928, an outbreak of the fungus in the eastern United States caused the loss of many millions of bushels of potatoes.

Losses due to potato blight are not widespread today, owing to the development of resistant varieties of potatoes and the constant application of chemical sprays. In some areas, as many as fifteen sprays are required. The crop protection experts tell us that there will never be another disaster with potatoes such as the 1846 famine. Let us hope they are right!

Ergot in Rye

Rye cereal grain is attacked by a fungus, *Claviceps purpurea,* which germinates in the spring to produce slender reddish-blue stalks with a pink globe at the top. The fungus produces millions of spores, which find their way by wind to rye flowers, Once there, they grow and produce a sweet secretion containing millions of a second type of spore. The exudation attracts insects, which spread the fungus very quickly. With cool, moist weather, the stage is set for a very rapid spread of infection. The infected kernels produce large purple cockspurs which are harvested along with the rye. The spores contain over 20 very potent alkaloids, which are very toxic to humans. The symptoms are frightening: intense pain in the abdomen, hallucinations, wild babbling, insanity, and finally death by convulsions. The ancient called the disease "holy fire" or "Heilige fever."

Ergotism has changed the course of history. The first known major outbreak occured in 857 AD. in the Rhine Valley and killed thousands of people. The weakened population of France could not resist the invasion from the north by the Scandinavians. Peter the Great, in 1722, led his forces against the Turks at Astrakhan and, like all armies at that time, lived off the land. The bounty included rye, again with ergot, and 20,000 soldiers died. Horses are also susceptible, and the Cossack cavalry was destroyed. The Russians never did capture the Dardanelles, even though they have tried six times since 1722. Ergotism continues to occur in all areas where rye is grown. In 1926 and 1927 an outbreak in Russia resulted in 10,000 casualties. The latest outbreak of any size occurred in Pont-St-Esprit, France in 1951. It was caused by ergot-infested rye flour that was added to wheat flour. The outbreak resulted in 200 cases of severe and damaging illness, 32 cases of insanity, and 4 deaths.

In the United States, historians have suggested that the Salem witch trials may have been instigated by ergot in rye. Certainly, some of the symptoms exhibited by the "witches" were similar to those produced by ergot in rye.

The story of ergot in rye is spectacular because the effect of the toxins is so visible. The decrease in yield of rye grain due to ergot infestation becomes minor. Adequate control methods are available to clean the grain so that the spores are not planted along with the grain. A simple flotation process involving a 30% salt solution will remove nearly all the spores from grain intended for seed or consumption. There are now very rigid government regulations concerning the content of ergot in rye.

There is a positive side to ergotism. The powerful toxins produced by the fungus have been isolated and found to have important medical applications. Some of them have been used to induce childbirth and abortion, and to prevent bleeding, high blood pressure, and headaches. One ergot alkaloid, discovered in 1938 by a chemist in Switzerland, turned out to be the hallucinogen lysergic acid diethylamide (LSD). The chemist tested it on himself and was terrified, apparently, by the sensation that he was floating outside his own body. LSD has since been used to treat schizophrenia and other mental disorders. For the purposes, LSD is synthesized rather than being obtained from ergot. However, ergot is still the main source for many of the other alkaloids.

The Wheat Rusts

The wheat plant probably originated thousands of years ago in the hills of Asia Minor, and wheat rust came soon after. Theophrastus of Greece, the "Father of Botany," described the rusts 22 centuries ago, and they are no less important today.

Wheat rust is a fungus disease that receives it name from the brown or rusty appearance of an infected field. Three main types of wheat rust are known—stem, leaf, and stripe rust—and many strains of each exist. The infection starts in the spring when, under warm, moist conditions, a spore carried on the wind reaches a wheat plant. The spore soon produces a fungus growth, which feeds on the wheat plant, and in a week or so the fungus produces thousands of spores. Each spore germinates upon reaching a wheat plant, and the cycle is repeated with frightening rapidity. As the wheat plant matures (with greatly reduced yields of grain), the rust parasite forms a different type of spore, called a teliospore. These can remain dormant all winter and germinate in the spring, producing yet another type of spore, the basidiospore. These spores fall on the young leaves of the barberry plant and produce male and female forms. These mate and produce asciospores, which are distributed in the wind and infect the young wheat plants. And so the cycle goes.

The alternate host for stem rust is the barberry plant. Leaf rust has two alternate hosts, the basilick plant and the meadow rue. Stripe rust has no known alternate host as yet. The existence of a sexual stage on the alternate host has awesome implications for wheat production. The millions of chance matings makes it possible for new races of wheat rust to develop at a rate that taxes the ability of the cereal breeder. Almost every new variety of rust-resistant wheat that has been developed painstakingly by the plant breeder in the past half century has been successfully parasitized by some virulent new form of rust.

Wheat is the most important cereal grain of commerce. It is grown around the world wherever the climate is appropriate, and specific varieties have been developed for each climatic area. The famed Marquis variety was introduced in the United States around 1900. It was immediately planted widely because of its yield and quality. Unfortunately, a rust race developed in 1916 and doomed the Marquis variety. The Marquis was replaced by a cross of Marquis and Kota called Ceres, but in 1935 the black stem rust (race 56) emerged and millions of acres of Ceres wheat were severely damaged. The Thatcher variety was developed but eventually had to be replaced by varieties that were more rust-resistant. In 1953, race 15B appeared, resulting in losses of up to 75% in durum wheat and 35% in bread wheat, Again, the geneticists developed new varieties, and they have since been able to keep one step ahead of the development of new rust strains. How long can this be expected to continue? There are no other established methods of control to date.

Other methods of control of wheat rust have been suggested. An obvious one is to eliminate the barberry plant. The first barberry eradication law was passed in France in 1660, and the first such law in the United States was enacted in 1726, but the barberry plant is still with us. The uprooting of millions of barberry bushes helped, but was no solution. Chemical sprays would certainly be effective in reducing the extent of rust damage, but fungicide treatment is not practical; it is simply too expensive. In 1938 the United States lost an estimated 100 million bushels of wheat to leaf rust. In 1953–1954, strain 15B cost the United States and Canada about 200 million bushels.

The Grain Smuts

The grain smuts are a group of fungi that attack wheat, corn, oats, and barley—in that order of importance. The flour made from infected grain looks dirty—hence the name. The fungus propagates by germinating in the soil in the spring. The fungus filaments enter the cereal plant and grow inside it, finally reaching the seeds. These seeds then become small containers, each holding millions of smut spores. With earlier agricultural methods, the smut spores were sown together with the grain and always took a percentage of the harvest. In order to sow cleaner grain, a series of chemical treatments was developed, starting around 1850 with copper solutions and progressing to formaldehyde and mercury salts, and, since 1950, hexachlorobenzene. These applications applied directly to the grain kernels have been effective and have saved billions of tons of food.

Chemical treatment of the seeds is effective for smut spores on the outside of the seeds, but some fungi infect the plant at the time of flowering and become embedded within the kernels themselves. These fungi germinate when the seed is planted. Chemical treatments strong enough to kill the embedded fungi would probably kill the seed as well. The new systemic chemicals that travel through the plant show promise of controlling this type of infestation, but conventional control is to rotate the crops. With increasing pressure for monoculture and continuous use of land, crop rotation and fallowing (leaving the land idle for one or more crop seasons) is becoming more difficult.

Plant pathologists have identified over 2,000 types of smut, and the number will undoubtedly increase. Scientific methods of control are making them less destructive than in prior years. Losses to smut in the winter-wheat belts in the United States are estimated at 3%. Fortunately, to date the bread wheats seem to be less susceptible. Before World War I, losses of wheat due to smut ran as high as 85% in some fields. At threshing time, the clouds of black smut were an explosive hazard to the threshing machines and very unpleasant for the workers. Smutty flour was not uncommon, but if it was too badly contaminated, bread made from the flour had a disgusting smell. An enterprising English miller attempted to utilize smutty flour by baking little cakes with molasses added to mask the color. He added a pungent spice to mask the unpleasant smell, and, lo and behold—gingerbread was created. Today, ginger snaps and gingerbread are not made from smutty flour, since there are very rigorous controls on the amount of smut allowed in commercial flour.

Smut consumes about 2% of the world's harvest of grain today. It has been kept to that low figure by chemical control and the development of resistant varieties. Apparently, smuts mutate very readily, and as we noted above, there are already over 2,000 of them. So the race goes on!

Corn Blight

In 1970, the United States lost 15% of its entire corn crop to an infestation of blight. It could have been a modern day disaster, but the technology that caused the disaster reacted so successfully and so quickly that the problem disappeared. The speed and extent of the successful reaction has no counterpart in history.

The corn blight was caused by a quirk in the production of hybrid corn. Hybrid corn is a source of pride in American agriculture, for it has been heralded as the greatest single breakthrough in agricultural science in this century. The story of hybrid corn began in 1917 in Connecticut when Donald F. Jones discovered the concept of hybrid vigor. He crossed an inbred line A with an inbred B to form an AB selection. Similarly, a line C was crossed with a line D to form CD. When the AB line was crossed with the CD line, the resultant ABCD progeny showed an increase in yield of 25%. It took 25 years to realize the commercial potential of this concept, but by World War II nearly all the field corn grown in the United States was hybrid corn.

Hybrid corn seed is produced by alternately planting six rows of AB and two rows of CD in the fields. The male pollen on the tassels of the AB rows is removed by taking the tassels at the top of the plant off by hand, and the ears of AB corn are then fertilized by the CD pollen, thereby ensuring the cross-pollination. The hand operation provided work for thousands of high school students but it was expensive.

In 1931, Marcus M. Rhoades discovered a corn plant with sterile tassels. The genetic factor for male sterility could be incorporated into the AB corn, and therefore eliminate the need for hand detasseling! Unfortunately, the progeny were also sterile. However, another male-sterile gene was discovered in Texas. In 1948, Jones reasoned that there must be a gene that restored fertility. He did, in fact, find such a gene, and he incorporated it into the CD corn. This meant that he could use a male sterility factor in the AB corn and a restorer gene in the CD corn and obtain fertile seed in the ABCD generation. The stage was set to produce hybrid corn without hand detasseling. The factor that caused male sterility was located in the cytoplasm of the cell, not the nucleus, and since it was found in Texas it was called the T strain. The T strain was soon widely incorporated into hybrid corn seed.

The use of T-cytoplasm also brought a liability. The fungus *Helminthosporium maydis* had always been present in corn fields but was considered a very minor pest. A report appeared in the Philippines in 1962 that T-cytoplasm corn was susceptible to *H. maydis*. The corn was tested in the United States with negative results, and the threat was dismissed as being due to different growing conditions in the Philippines. Unfortunately, a new race of *H. maydis* (race T) appeared in 1970, and all corn with T-cytoplasm was susceptible. The blight appeared first in 1970 in Florida and then moved north and west with frightening rapidity. Over 500 million bushels of corn were lost. Worried corn specialists feared that in 1971 the United States might lose almost its entire corn crop.

The scientists soon located the problem, and a crash program was started to eliminate seed corn carrying the T-factor. Seed grain was produced in the off-season in Mexico, and in as many other areas as possible, so that enough resistant seed would be available for the 1971 season. It was not possible to shift such a large industry so fast, so resistant seed was allocated to the more susceptible southern areas and T-strain corn was, of necessity, diverted north. Fortunately, the weather in 1971 was unfavorable to the spread of the blight and disaster was averted. In 1971, sufficient resistant seed was produced for the whole 1972 crop and the disaster was eliminated.

Hybrid corn is again being produced in the United States by hand detasseling. This was the first time that cytoplasm had been shown to be an important aspect of diversity. A new dimension had appeared in plant breeding. The technology that produced the potential for disaster removed it. With an interesting form of hindsight, it was perhaps fortunate that the near disaster occurred with hybrid corn. The ability of the breeders to form new combinations enabled them to eliminate the problem in one year. If the susceptibility was due to a factor in an open-pollinated variety, the solution may have been much more difficult.

Interestingly, the corn smut spores themselves are edible. In Mexico, immature corn smut spores are wrapped in pastry, and deep fried.

Coffee Rust

The story of the coffee culture is a fascinating example of how a disease can change a country's way of life. The coffee plant probably originated in Ethiopia. Turkish coffeehouses existed in the fifteenth century. The Europeans became very interested in this new drink and attempted to smuggle seeds for many years with no success. However, by 1600 some seeds had been taken to India. When the Dutch came to India as traders, they found coffee trees in the hill country. The Dutch planted coffee trees in Ceylon and Java, and the coffee culture was then spread through the Caribbean Islands, Brazil, and Mexico.

Britain took over Ceylon in 1797 and in 1825 reestablished the coffee plantations there. Coffeehouses had been established in Britain in 1652 and were very popular for over a century. It was to cater to the coffeehouse trade that Britain reestablished the coffee plantings in Ceylon. The boom was on! Exports of coffee increased from almost zero in 1830 to 100 million pounds in 1870. Planters grew rich beyond their wildest dreams, and then disaster struck. The coffee rust *Hemileia vastatrix,* which was similar to grain rust, spread through the coffee plantations. The spores germinate on the coffee leaves in warm, humid weather and produce more spores in a matter of weeks. The wind-blown spores soon devastated entire plantations. The weakened trees withered and died. By 1892, coffee exports from Ceylon, the world's leading exporter, had dwindled to zero. There was no cure for the coffee rust in Ceylon, and there was only one alternative—switch to tea. Starting in 1875, the dying coffee plantations were replanted with tea. This was a stupendous operation, one that has no parallel in history. In the meantime the British had been importing tea, since it was cheaper than coffee. Britain promoted its tea trade throughout the Empire and became a tea-drinking country. Ceylon switched from a coffee to a tea economy.

Control of the coffee rusts centers around growing the trees in the shade. Indifferent success was obtained with a number of chemical sprays. The real answer is the development of resistant varieties. Fortunately, such attempts have been partially successful and at least one variety of *Coffee arabica* seems to have resistance. Other hardy varieties are derived from *Coffee robusta,* which has quite a different taste. Nearly all of the robusta in commerce today is grown in Africa. Most experts consider robusta to be much inferior to arabica.

Ninety percent of the world's coffee is grown in the Western hemisphere. Rust spores have spread east to the Philippines and to the west coast of Africa. If they do jump the 2,000 miles of ocean to the New World, either by ship or airplane, they will surely devastate the plantations of arabica in Brazil and Columbia. The fungus spores did reach Puerto Rico once in 1903, but an alert inspector destroyed the plant before the spores could become established. If the disease does spread to the New World, robusta may be the only type of coffee available.

Banana Diseases

The banana trade in the United States offers an interesting tale of agriculture, science, and politics. The banana fruit itself is estimated to be a million years old; some even say that the "apple" in the Garden of Eden was a banana.

It became apparent to entrepreneurs after the American Civil War that opportunities existed in the banana trade. Hundreds of companies were established during the nineteenth century to produce and ship bananas to the United States. Business was so good that around 1900, over 800,000 tons of bananas were being shipped annually to the United States. The United Fruit Company emerged as the giant in the trade and in the 1950s shipped over 2 million tons of bananas per year to the United States and Europe.

In Central America, the pressure on the land to produce bananas and the subsequent monoculture led to the first serious outbreak of Panama wilt in 1915, and production dropped. Panama wilt is a fungus disease (*Fusarium oxysporum cubesse*) that remains in the soil until it reaches a banana plant. It grows within the plant and soon kills the plant. Enough virgin land was brought into production to restore the crop. However, in 1930 a second wave of Panama wilt reduced production by about 50%. Flooding the land for six months was found to be a remedy, but that works only if enough land and enough water are available.

Around 1935, a new scourge attacked the banana plants—sigatoka disease, caused by the fungus *Cercospora musae*. This fungus germinates from spores on the banana leaf and soon destroys the plant. It appeared in the Caribbean, first in Trinidad, and then spread rapidly to Jamaica, Honduras, and Guatemala. Production dropped rapidly, and a feverish search for a cure began. A cure of sorts was found in the form of sprays of a well-known chemical containing copper sulfate—Bordeaux mixture. Another type of mineral oil spray gives some protection. These two banana-plant diseases have been kept under partial control in recent years by a combination of chemical sprays and land flooding, but it is an uneasy truce. A more virulent form of either disease could reappear at any time. Eventual control will probably depend on the production of resistant varieties, but progress in this area is very slow.

Miscellaneous

The diseases discussed above were chosen because of their spectacular effect or sheer magnitude, but there have been many other, lesser-known plant diseases. The bacterium causing the fire blight of pears devastated the pear orchards around 1900, and even today, pears cannot be grown in some areas. A bacterium causes crown gall (a type of cancer in trees), for which no control other than eradication is known, The citrus canker threatened to eliminate the citrus industry in the United States, and a frantic total eradication program was started in Florida and Mississippi. If a single infected tree was found, the entire orchard was destroyed. The operation succeeded and the disease never reached Arizona and California. In 1984, another outbreak of citrus canker occurred. It was controlled after the destruction of millions of citrus seedlings. The invasion of the Mediterranean fruit fly threatened to destroy the entire Florida citrus industry in the 1930s. A large-scale spraying program eliminated the insect. Another outbreak of the Mediterranean fruit fly occurred in Florida in 1956. Three years and $12 million later, the insect was eradicated. The insect made its fourth appearance in Florida in 1981 and was promptly controlled. Three outbreaks have occurred in California. The most recent was in 1980–1981 with a cost of $100 million. The California fruit and vegetable crop susceptible to the fruit fly is estimated at $15 billion, so $25 million for control measures seemed like a good expenditure. In 1983, California reported an outbreak of a similar insect, the Oriental Fruit Fly, and yet another control program was organized. The struggle goes on.

If we study the history of trees, we may be glad that we do not eat them. The American chestnut was a wonderful tree with a trunk diameter up to 4 feet and a height of 100 feet. It provided superb lumber as well as chestnuts for the wild turkeys. It grew from Maine to Michigan and south to Louisiana, constituting over 25% of the total tree population in this country. In 1904, a tree in the New York Botanical Gardens began to die of a fungus disease. The disease blocked the tracheae of the tree and spread very rapidly. By 1935, there were no living chestnut trees. After death, the trees are usually invaded by wood borers, which produce the characteristic pattern of wormy chestnut now available only for picture frames. The disease crossed the ocean and is now eliminating the chestnut trees of Italy and southern Europe.

The American elm is a gracious and imposing shade tree. In 1930, two sick trees were discovered in Cleveland and Cincinnati. A fungus disease was carried in elm logs imported from Holland destined for furniture veneer, and it spread rapidly. There is no cure to date, and the American elms are doomed. The total cost of removal of the infected trees is estimated at many billions of dollars. Oak trees are also being infected by a newly discovered fungus disease, but the rate of spread is much slower. Maple trees showed a mysterous decline during the drought in the 1950s, but, fortunately, this malady seems to be slowing down. Let us hope that some of our important food crops do not go the way of the chestnut and the elm.

Theoretical Approaches

The spectacular successes of the Green Revolution with wheat and rice have been criticized on the grounds that the many millions of acres planted to genetically similar varieties represented potential disaster. Modern plant breeders realize this and have labored to broaden the genetic base. With every passing year, the potential for disaster decreases as new varieties are introduced. The famous IR-8 dwarf variety of rice released in 1966 launched the Green Revolution in rice, but it was susceptible to many diseases. It was replaced with IR-36 derived from cross-breeding parents from six countries. It is resistant to eight major diseases in rice. The latest major variety IR-58 released in 1982 was only possible through the breeding efforts of IRRI (see Chapter 7), supplemented by the genetic diversity available in a very large collection of different rice varieties. IR-58 will probably last only 5 or 6 years before it too becomes susceptible to a disease mutant, so another variety must be ready to take its place.

The National Academy of Sciences in the United States has surveyed all the major American crops to determine the degree of uniformity and vulnerability among them. For example, 96% of the pea crop is planted to only two types, and 95% of the peanut crop is composed of nine varieties. But in the case of wheat, the largest two varieties constitute only 25% of the crop. The degree of uniformity desired in the American food supply contributes to a narrow genetic base. The genetic uniformity may even extend to several varieties. For example, the same dwarfing gene is used for several varieties of wheat. Similarly, with rice. The stringless gene in beans is similar in several varieties. The gene in tomatoes that makes the plant set all its fruit at the same time is incorporated into a number of varieties. This type of genetic character may or may not make the plant susceptible to a new race of parasites. As we move from chemical control to genetic resistance, this becomes an interesting question.

The ability to keep developing varieties with resistance to diseases depends primarily on the availability of a very diverse gene pool. These gene pools are provided mainly by collections of plants and animals that have developed and diverged down through the millenia. The United States has a National Plant Germplasm System that holds over 450,000 samples of plants. Similar systems are being established around the world under the aegis of the Consultative Group or International Agricultural Research (see Chapter 7), but the criticism is mounting that valuable plant resources are being lost at an increasing rate. It is absolutely vital that the diverse gene pools be maintained if this world hopes to feed itself.

THE CHANGING CLIMATE

Recent increases in population have triggered a great deal of discussion, both scientific and emotional, on the capacity of the earth to provide enough food. We are not yet in an era in which significant quantities of synthetic food are being produced, so we have to depend on agriculture. Conventional agriculture is so finely tuned to climate that any change could have devastating effects on food production. Changes as small as a drop of 1°C in average temperature can have a large effect on food production. A drop in average temperature of 2°C would eliminate wheat production above the thirty-eighth parallel—that is, wheat production in all of Canada and all of the United States north of Kansas. An average drop of 3°C could produce another ice age. Conversely, an average rise of 3°C would melt polar ice and flood most of our coastal cities.

Most long-range climatologists agree that the earth's temperature is a delicate equilibrium between energy received from the sun and energy dissipated by the earth into space. Many complex forces operate to maintain this equilibrium, and the role of modern long-range climatologists is to attempt to understand these complex phenomena. Progress in this field is vital if we are to develop climate forecasting systems that will enable agriculturalists to respond to changing patterns in order to avert future famines.

Famines are not new to this planet. Historians have clearly identified 49 famines between the years 975 and 1804. Famines have been correlated with cold periods in the earth's history. However, the influence of the colder climate was less important in the past because there was less pressure on the land to produce food then. With our greatly increased population and consequent greater dependence on the available land, future cold periods become much more important. The forecasters tell us that we have now entered another cold period. It may be that the short-term trend will be colder, but the long-term trend may be warmer.

The Cooling Trend

The methods used by the long-range forecasters make a fascinating tale. They include analyses of the type and quantity of pollen in bogs, the deposition of lead as dust in glaciers, the variations in tree growth rings, the changing ratios of isotopes of oxygen in earth cores, and the settling of atmospheric dust.

It is possible to estimate the average temperature of Iceland by measuring the duration and extent of the ice floes covering the land and surrounding water. The position of the ice and the average temperature have been measured since 1890. The Vikings kept excellent records of the duration and extent of the ice for the last thousand years. Thus, by calculating the relation between average termperature and ice position and projecting backward in time, the average temperature could be calculated back to 900 AD. The average temperature for Iceland from

1000 to 1974 AD is plotted in Fig. 15.1. The interpretations are both fascinating and scary. The Vikings colonized Greenland in the tenth century, and apparently that land was aptly named. But by the fifteenth century, the colonies had been frozen out. There was a brief warm spell around 1400 AD and a "little ice age" between 1550 and 1900. The earth started to warm up about 1890, and this trend continued until 1945. Since 1945 the earth has been cooling rapidly. The cooling trend levelled off between 1965 and 1975 and then resumed its downward trend. There is no evidence to date that the downward trend has stopped. The cooling trend refers only to the Northern hemisphere, since there is some suggestion that the Southern hemisphere may be constant in temperature or even warming slightly. Unfortunately, there are few recording stations in Southern hemisphere.

The warming trend from 1890 to 1945 was only 3°C and the cooling trend since 1945 is only 2.8°C. Yet the results of the cooling trend are dramatic. Hay yields in Iceland were reduced by 25%. The growing season in England is 2 weeks shorter. Icelandic fishing fleets can only fish to the south now because of drifting ice. In 1979, all five of the Great Lakes in Canada froze over for the first time in memory. In 1983, 50 Soviet ships were trapped in the ice near the Bering Strait on the Northern Sea Route, due to the coldest winter in 100 years. The drop in temperature since 1945 is believed to be the longest unbroken trend downward in hundreds of years.

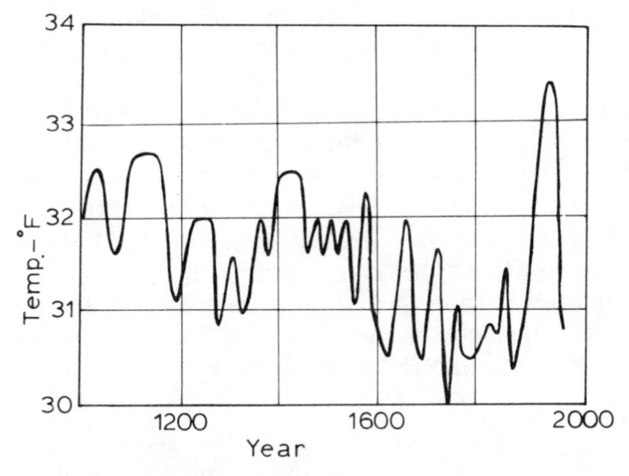

Fig. 15.1. A 900 year history of the temperature of Iceland as estimated from the position of the ice.

The 300- to-400-year cycles, important as they may be in themselves, become even more ominous when superimposed on a 100,000-year cycle. Researchers have been able to plot the average temperature of the earth over the long periods by studying the ratio of oxygen isotopes in shells from sea-bed cores in the Caribbean. Apparently, the oxygen ratio is affected by the water temperature. When the temperature is plotted against time for 700,000 years, a 100,000-year cycle emerges. Apparently, we are now at the temperature peak of one cycle; that is, we are experiencing the warmest temperatures that have occurred in the last 100,000 years. If we super-imposed the 300- to-400-year cycle described above and the 100,000-year cycle, it is obvious that the greatest increase in population has occurred at a time when we are enjoying the most advantageous climate in the last 100,000 years. If we are concerned about the food supply in this period, how can we hope to feed the world when the climate gets worse? Yet the population growth rolls along!

Broader Implications

The above prospect is chilling enough but the cooling trend has many other implications. The cooling of the earth results in an increase in the polar ice caps. The cold air above the ice caps interferes with the air currents moving around the globe. The large-scale circulation in the atmosphere is influenced by the temperature difference between the equator and the poles. In simplified terms, warm air rises at the equator and dumps much of its moisture on equatorial rainy belts. The dry air moves up and out toward each pole, then descends to form the world's major deserts. Some of the air moves back toward the equator and forms the trade winds. The remainder moves further north or south and forms the westerlies at low altitudes and the jet stream at higher altitudes. The cold air over the poles pushes both types of winds further south, and this is probably one reason why the Sahara Desert is moving south.

The six-year drought in the Sahel, a band south of the Sahara Desert comprising parts of the countries of Senegal, Mali, Mauritania, Upper Volta, Niger, and Chad, has caused untold misery and famine to the people of that area. The climate change may be only one factor in the enlargement of the desert in the Sahel area. This enlargement may also be a classic example of "The Tragedy of the Commons."[1] When the

[1] *The Tragedy of the Commons* is a classic environment film in which a town common was shown to be able to support, for grazing purposes, a given number of cattle. Since the town common belonged to all the townspeople, some insisted on attempting to graze more cattle. This resulted in destruction of the grass cover and the town common could not support any cattle.

French colonial era ended, the old tribal systems began to break down and the nomadic population doubled. More important, since animals were considered a source of wealth, the cattle population tripled. In times of plentiful rainfall, adequate fodder was available, but when the drought started the land could not support the animal population and the system collapsed. The Sahara Desert started to move south at a much faster rate. This movement could be slowed down even in drought years by appropriate land management. This was pointed out by American observers who, while studying satellite photos, noticed a green pentagon-shaped area in the desert. A team of investigators found that it was a 250,000-acre ranch partitioned into five areas. Each area was grazed once in five years. The ranch had been started in 1968 after the onset of the drought, and the only difference between the green area and the desert was a barbed wire fence. In 1982, drought returned to the Sahel, and the local population again is in desperate straits. The reclamation of the desert is a tremendous task, but some progress is being made. For example, Algeria has started the most ambitious project of the century—the building of a 950-mile tree barrier across the entire country, approximately 200 miles south of the Mediterranean Sea. Composed of eucalyptus and Aleppo pine, the green belt will vary from 3½ to 15 miles wide and is expected to take 20 years to complete at a minimum cost of $100 million. Hopefully, it will help return Algeria to its agricultural productivity of 3,000 years ago. Yet Algeria is a small part of the land menaced by the Sahara Desert.

The winds that are pushed south by the polar air masses have important implications for other areas. They form the monsoons that bring life-giving rains to India. Part of this rain has been falling in the ocean, and drought is increasing in India. The same conditions are affecting the grain-producing steppes in part of the USSR. In North America, the West is becoming colder and wetter and parts of Kansas have had rainfalls more severe than any of the inhabitants can remember. Part of the grain-growing area in Canada's northwest had a snowfall in August. The cooler air from the west that is moving east and being warmed on the way has contributed to warmer winters in the eastern United States.

A drop in temperature usually, but not always, signifies a drop in agricultural production. It depends on whether or not temperature is the factor that is limiting growth. Apparently, with corn in Missouri it is not, because it is estimated that a drop in summer temperature of 1°C and a 10% increase in rainfall will increase corn production by 20%. The increased rainfall in the West might well increase grain production in areas where rainfall is the limiting factor. This will mean a return to the situation occurring at the time of the "forty-niner" gold rush in the sense that a hazard of crossing the plains was the height of the grass; it was head-high, and one could lose sight of the main party. These areas of the plains are practically desert today. The decrease in grass alone could have accounted for a 50–75% drop in the numbers of buffalo, aside from the overhunting. The cooling trend combined with the increase in rainfall might not harm the United States, but it could be disastrous for areas further north around the world.

The Greenhouse Effect

One may well ask, "Why is the earth cooling off?" Apparently, the earth's climate is a balance between the "greenhouse effect" and the "particle effect." The greenhouse effect refers to the trapping of the sun's energy by carbon dioxide. The energy from the sun is in the form of short, visible, wavelengths which can pass freely through carbon dioxide. When this energy reaches the earth's surface, it is converted into heat and reradiated back as infrared rays. Carbon dioxide molecules absorb a portion of the infrared rays, thereby trapping the energy and creating a rise in temperature. The concentration of carbon dioxide in the atmosphere from volcanoes and use of fossil fuels rose from 315 to 339 ppm from 1958 to 1979. The rate of increase is also increasing, from 0.7 ppm per year in 1959 to 1.7 ppm per year in 1978. Manabe and his coworkers at Princeton University, have estimated the temperature increase resulting from a doubling of the concentration of carbon dioxide in the atmosphere. The temperature rise at the tropics would be less than 2°C. At a latitude of 40°N (roughly equivalent to Los Angeles and Tokyo) the rise would be 3°C. At 50°N (Paris and Vancouver) the rise would be over 4°C. At 76°N (above the Arctic Circle) the increase would be 7°C. The actual warming trend observed to date has not risen above the level of the yearly fluctuations in temperature, which is about 0.2°C. If the warming trend due to the increase in carbon dioxide is real, an unmistakable warming trend should emerge in the 1990's.

Other compounds besides carbon dioxide may contribute to the greenhouse effect. Nitrous oxide from degradation of organic matter and agricultural fertilizers, methane from biological degradation and fermentations, and chlorofluorocarbons from industrial activities may contribute as much as 40% to the greenhouse effect as compared to 60% for carbon dioxide. Water vapor also contributes to the greenhouse effect, but is controlled by the climate in a complex and little understood feedback system.

The Particle Effect

An accumulation of particles in the upper atmosphere can reflect the sun's rays back into space, reducing the energy reaching the earth's surface, and thereby producing a cooling effect. It is possible to estimate the concentration of particles in the atmosphere back through history by studying the dust deposits in the ice in the polar areas. Historically, the greatest contributor has been volcanic activity. Volcanoes can produce tremendous quantities of dust in a very short time. The eruption of Tamboro in the Philippines in 1815 produced so much dust that as far as 500 kilometers away there was total darkness for three days. The world temperature fell about 1.1°C below normal, and 1816 became known as the "poverty year" and the "year without a summer." Tamboro was the biggest volcanic eruption on record, and hurled an estimated 20 cubic miles of solids into the atmosphere. The eruption of Laki in Iceland and Asama in Japan, both in 1793, combined to make the following three years among the coldest on record in the Northern Hemisphere. The eruption of Krakatau in the Philippines, in 1883, blew an estimated four cubic miles of solids into the atmosphere and plunged the island into total darkness. It also created a tidal wave of 120 feet and killed 30,000 people. The recent eruption of El Chichon in 1983 was much smaller, but is still expected to influence the climate in North America. By way of comparison, the eruption of Mount St. Helens in the U.S. in 1980 produced a dust cloud about one-twentieth of El Chichon and had no measurable climatic effect. Volcanic activity around the world was relatively low from 1920 to 1955 but seems to be increasing lately. The dust from volcanic eruptions produces spectacularly colored sunsets around the world, but they bode ill for agricultural production.

Another natural source of dust in the atmosphere is salt particles produced by the evaporation of salt water. Another may be the natural smog produced by forests. It is assumed that natural forces are the major contributors of the particles in the atmosphere, but there is evidence that man's activities are contributing more and more. The dust pall resulting from overgrazing by nomadic tribes may have contributed to the spread of the Thar Desert in northwest India and Pakistan. The dust from barren land increases the turbidity of the lower atmosphere, thereby reducing the upward currents. This reduces the cloudiness and increases the aridity. This, of course, is a local effect, but the dust from extensive agriculture probably does increase the number of particles in the air. The "slash and burn" methods of land clearance in the tropics undoubtedly contribute a great deal of smoke. Increased industrialization certainly produces more smoke particles. Whatever the source of particles, there is solid evidence that the intensity of sunlight reaching the earth's surface increased from 1910 to 1930 and then started down. The temperature in the Northern Hemisphere increased from 1915 to 1945 and then started down. The lag from 1930 to 1945 was probably due to the greenhouse effect. Some workers have attempted to correlate the intensity of sunlight reaching the earth to sunspot activity on the sun. Presumably, sunspot activity is beyond human control. It seems that the particle and greenhouse effects are more logical explanations of climate and are open to some modification.

The Warming Trend

There was considerable scientific concern in the 1970s over the agricultural implications of the cooling trend in the Northern Hemisphere, but lately the consensus is that we should be concerned in the longer term about the warming trend. The greenhouse effect, due primarily to carbon dioxide in the atmosphere, is still a theoretical concept, but is has been bolstered by other considerations. For example, it has been suggested that the very low overall increase in temperature has been due to melting of the polar icecaps. The West Antarctic ice shelves are producing icebergs at the rate of 500 km^3 per year. This should result in a rise of 1.5 mm per year in sea level, and this is what is being observed in worldwide averages obtained from tidal gauges. This added liquid has been correlated with a change in the inertia of the earth resulting in a measurable reduction in the earth's rate of rotation. Supporters of the greenhouse theory point to the recent space probes correlating temperature and carbon dioxide levels. The atmosphere of Venus is 97% CO_2 and its temperature is 527°C. Mars has very little CO_2 and its temperature is −53°C. Earth has 0.03% CO_2 and more water vapor than Venus or Mars and its temperature is 15°C. Admittedly, the evidence is very speculative, but the consensus seems to be that the average global temperature of the earth is going to rise.

The implications of the temperature rise accompanied by the melting of polar ice are obvious for the coastal cities. Many coastal cities would be flooded and others would be more susceptable to storm damage due to the higher water level. If all the ice above sea level in the West Antarctic ice sheet were to melt, it would raise the sea level by 6 meters. Half of the state of Florida would be under water. This level was actually reached during the last interglacial period (120,000 years ago) as judged by the level of the coral reefs.

The implication of the temperature rise for agricultural production are less obvious. Most climatologists, who are brave enough to predict, call for a 2–3°C rise early in the next century. This will change the world patterns of rainfall. There actually was a time between 4500 and 8000 years ago when the earth was several degrees warmer. This together with several anomalies in our historical temperature patterns and mathematical models, give some idea as to what could happen. The United

States would become drier and hotter, and would have difficulty maintaining its current production of half the world's maize. In 1978, Russia produced nearly one-third of the world's wheat, potatoes, and barley, and these crops would face drier conditions. China and India produced over half of the world's rice and the longer, hotter and wetter seasons would be to their advantage. The increased carbon dioxide in the atmosphere would make plants grow faster and possibly be less subject to water stress. The effects of more or less water, higher or lower temperatures, longer or shorter growing seasons will change with each crop, thus any predictions are precarious. Also, the type of crop grown in a particular locality may change. Possibly, the plant breeders can adapt the crops to the changing conditions. One thing, however, is certain—the patterns of world trade in foods will certainly change.

The concern over the potential rise in temperature has resulted in a number of major reports. The Environment Protection Agency (EPA) sponsored in 1983 a report entitled "Can We Delay a Greenhouse Warming." The authors developed a "best guess" of future energy patterns. They estimated that atmospheric CO_2 level would reach 590 ppm, or double preindustrial levels, by 2060, and a 2°C temperature rise would occur around 2040. They predicted the effect of some energy policies on the changes in the date of the 2°C warming. A total ban on shale oil and syn-fuels would delay the date by 5 years. A total ban on coal would push the date on by 15 years. A total ban on both coal and shale would delay the rise by 25 years.

The rate of production of CO_2 is rising at about 2–3% per year due to the increase in industrialization. The CO_2 in the atmosphere dissolves very slowly in the ocean, which is estimated to contain as much as 60 times the CO_2 in the atmosphere. Because the rate of dissolution of CO_2 into the deeper layers of the ocean is so slow, it could take as much as 100 years to reach equilibrium. The rate of production of CO_2 is obviously greater than the rate of solution in the ocean, and this may be the major reason for the increase in the amount in the atmosphere.

Fossil fuels are the major sources of CO_2 in the atmosphere. The clearing of forests undoubtedly contributed in the past because green plant tissue removed CO_2 from the atmosphere by photosynthesis. Removal of the trees allowed the CO_2 to accumulate, but, at present, present reforestation is estimated to equal deforestation. Of the fossil fuels, gas and petroleum are of limited supply so the major contributor will be coal. Russia, United States, and China have 83% of the world's reserves of coal so obviously any agreement to reduce the rate of CO_2 buildup must depend on a political agreement between them. If the greenhouse theory does hold up, one can make a good case for the development of alternate sources of energy (see Chapter 15). Nuclear and solar power may yet come into their own. One thing, however, is certain. The forces contributing to the CO_2 buildup are already in place and are unlikely to change. Future research will only serve to identify more clearly the winners and the losers. The political aspects are awesome.

Other Approaches to Climate Change

The increasing concern of the world's governments with greater food supplies has led to many schemes—some grandiose, other quite practical. The Aswan Dam in Egypt, which provides power and irrigation to 2½ million acres is an accomplished and practical feat of engineering. It has probably modified the climate of the Nile valley to some extent, probably for the good. The large dams currently being planned for India, Pakistan, China, Russia, Africa, South America, and many other areas fall in this category. If the large dams involve a change in direction of flow of the rivers, we can anticipate other climate changes. For example, it is difficult to forecast the effect of the current project in Quebec in which waters from three rivers that normally flow into James Bay are being diverted south into the St. Lawrence system. Another example is the suggested southward diversion of the MacKenzie River in Canada; it normally flows into the Arctic Ocean. The diverted water would flow east to the Atlantic Coast and south as far as Mexico. The project was estimated to cost over $100 billion and would involve a mind-boggling degree of international cooperation. The proposal to dam the Straits of Siberia to prevent the cold Arctic waters from flowing south would probably warm the climate of Siberia. It might change other areas as well. The proposal to extract heat from the Gulf Stream as a source of energy has implications for changes in climate if an appreciable amount (10%?) of the energy of the Gulf Stream is removed. The current status of cloud seeding has implications for climate change—to whose advantage, one might ask. We can summarize some of the more grandiose schemes by saying that a great deal of thought must be given to all of them before we create more problems than we solve.

BIBLIOGRAPHY

ANON. 1972. Genetic Vulnerability of Major Crops. National Academy of Sciences, Washington, DC.

ANON. 1976. Climate and Food. National Research Council, Washington, DC.

ANON. 1982. Carbon Dioxide and Climate: A Second Assessment. National Research Council, Washington, DC.

ANON. 1983. Changing Climate. National Research Council, Washington, DC.

BRYSON, R. A., AND MURRAY, T. J. 1977. Climates of Hunger. University Wisconsin Press, Madison, Wisconsin.

CAREFOOT, G. L., AND SPROTT, E. R. 1967. Famine on the wind. Rand McNally, New York.

REVELLE, R. 1982. Carbon dioxide and world climate. *Sci. Am.* **247**(2),35–43.

SEIDEL, S., AND KEYES, D. 1983. Can We Play a Greenhouse Warming. U.S. Environmental Protection Agency, Washington, DC.

Index

M

N

O

P